Introduction to Ethology

Introduction to Ethology

Klaus Immelmann
University of Bielefeld
Bielefeld, West Germany

Translated from German by

Erich Klinghammer
Purdue University
Lafayette, Indiana

PLENUM PRESS · NEW YORK AND LONDON

Library of Congress Cataloging in Publication Data

Immelmann, Klaus.
Introduction to ethology.

Translation of Einführung in die Verhaltensforschung.
Includes bibliographical references and indexes.
1. Animals, Habits and behavior of. I. Title.
QL751.I5213 591.51 80-15721
ISBN 0-306-40489-3

First Printing – November 1980
Second Printing – September 1983

This volume is a translation of the first edition of *Einführung in die Verhaltensforschung,* published by Paul Parey Verlag, Berlin and Hamburg, in 1976. It contains additional material that was included in the second German edition of the work, which was published in 1979 concurrently with the preparation of this translation.

A Division of Plenum Publishing Corporation
227 West 17th Street, New York, N.Y. 10011

Printed in the United States of America

Preface

Ethology, the study of the biology of behavior, has grown tremendously during the last few decades. The large number of accumulated facts is difficult to survey, understanding and an appreciation of the ethological approach to the study of behavior have grown, and the number of attempts at holistic explanations for certain behavioral phenomena has increased. Because of this development it has become more difficult to gain an overview of the field, to keep up with new developments, and to update the subject matter by the inclusion of new facts in the proper place. The nonspecialist is unable to evaluate the more general statements in the popular literature, especially when such works are aimed at a broader audience.

Hence, this book has a dual purpose: (1) to lend some order to the dizzying array of information and thus simplify inquiry into ethology; and (2) to present relevant facts and knowledge that will help the reader confronted with numerous studies and articles in the ethological literature.

It is always difficult to select material for such a survey and to arrange it in some logical manner. Hence, this volume offers but one of many possible ways to present this material. This book is based on a course of lectures of the same title that I have offered during the past 12 years at the universities of Braunschweig, Berlin, and Bielefeld in West Germany. These lectures were continually modified in response to questions that indicated what knowledge in which particular places would enhance the understanding of certain relationships. Thus, this book can be regarded in its organization as a sort of stepping-stone on the path to a solution. This, I hope, will be the clearest and simplest way to present the relationships among facts, hypotheses, and theories.

For reasons of space, limitations existed on the selection of material. This is why the physiological basis of behavior, which is generally offered in texts and surveys in physiology, is treated less comprehensively than the area of "classical" and comparative ethology. However, even here focal points had to be chosen; thus, the treatment of topics in the currently popular style of teaching by illustration and example is somewhat uneven. As a result, some areas, such as aggression, where general regularities of behavior and methods of study can be clearly demonstrated, have been presented in more detail than have some others. The extensive area of learning has been presented in comparatively condensed form, since numerous texts on the subject are readily available.

Since this is an introduction that is primarily designed to provide the means for a more intensive involvement with the field, the author has not participated in the sometimes controversial discussions about basic concepts in the various "schools" of ethology. These are pointed out in those instances where certain kinds of explanations and theories are still being debated.

With respect to humans it must certainly be accepted that some characteristics of human behavior developed as the result of the evolutionary process, and are hence comparable to corresponding aspects of the behavior of animals. Hence, ethology is in a position to point out a certain regularity that in some way is also applicable to humans. Such applications must, however, be made with great care since insufficiently supported comparisons could, as has been shown repeatedly in the past, only harm ethology in its attempts to make more general statements about behavior, and thus to aid in a better understanding of human behavior as well.

Ethology is a biological science. Each behavioral characteristic of an animal can be understood and interpreted properly only when its biological significance, i.e., what its importance is for the particular species in its natural environment, is known. This has not always been the case in the history of the study of animal behavior to the extent that is necessary. The continued reference to such ecological relationships, and to adaptations that have developed during the course of evolution, is an additional aim of this book.

Since an introduction of this kind is also directed to readers interested in ethology who may not have any special biological background, it was necessary to explain certain basic concepts that are important for an understanding of the biological significance of a particular behavior. These are found in the text and in footnotes. This also applies to special terms that are not properly a part of ethology.

The scientific names of the animal species cited are given only when a proper common name is not available.

I am indebted to the illustrator, Mr. Klaus Weigel, of the Faculty of Biology of the University of Bielefeld, for the preparation of the drawings, which he carried out with great care and understanding. Thanks are also due my colleagues and assistants in Bielefeld for their stimulating ideas and discussions during the writing of this book.

Klaus Immelmann

Bielefeld, March 1976

This English edition contains additional material that was included in the second German edition, which appeared while the English translation was in progress. Two new chapters, one on domestication and one dealing with ethology and human behavior, were added. These chapters deal with research that may lead to a better understanding of human behavior as well. However, the facts that have been collected so far permit only limited conclusions to date. For this reason, these chapters are of a more speculative nature than the previous chapters. Indeed, the last chapter was added only in response to requests for its inclusion, although there has been no change in the views expressed on this subject in the introduction to the first edition.

Klaus Immelmann

Bielefeld, July 1980

Contents

Introduction to Ethology

1

Aims, Methods, and Areas of Ethology

1.1. Definitions of Concepts

The goal of ethology is the investigation of behavior with the methods used in the natural sciences. However, to define the concept of *behavior* is not as simple as it may seem, and the use of this term in the literature is not consistent. Here is a problem from which ethology has suffered in particular: the difficulty of clear definitions. This is due to the variety of the phenomena studied, which in turn has led to some premature interpretations.

As a rule, the word BEHAVIOR is used in a very broad sense in ethology. It refers to movement patterns, vocalizations, and body postures of an animal, as well as to all externally recognizable changes that serve in reciprocal communication and that can release behavior patterns in another animal, including changes in coloration or the release of odors. Hence, behavior is not restricted to movements, and an animal that appears to be quite inactive may "behave" in this use of the word. For example, a male antelope standing motionless on a termite hill indicates his ownership of a specific area as its territory, and a female butterfly releasing an odor (pheromone) with which to attract a male "behaves" according to this broad definition.

The term ETHOLOGY (from the Greek *ethos*—habit, convention), as is used today for the biological study of behavior, is quite old, but it was originally not used as it is today. The term first appeared in the middle of the 18th century in publications of the French Academy of Sciences, where it was applied then, as later in the 19th century, to the description of life-styles in general, i.e., what is at present considered under the heading of the ecology or general biology of an animal species. The application of the term to the study of behavior in a more limited sense was introduced in 1950 by N. Tinbergen.

In the early years of ethology the synonym *animal psychology* (German: *Tierpsychologie*) was widely used. Today, with the increasing physiological emphasis in ethology, the term is no longer as appropriate and has come into disuse as a result. However, the term ANIMAL PSYCHOLOGY is still applied in this more limited sense by ethologists when it refers to individual, subjective (to the degree that this can be investigated by objective criteria), and disease-related phenomena in the behavior of animals. Important psychological knowledge has been gained from animals in the zoo and circus. In the United

States and Great Britain the subject matter of ethology is usually referred to by the term ANIMAL BEHAVIOR in a more general sense. The term COMPARATIVE PSYCHOLOGY refers to the comparison of species with respect to categories of behavior traditionally of interest to American psychologists, such as learning, motivation, and others that were usually investigated with the animal as a model for human behavior.

1.2. Descriptive Ethology

Ethological analysis consists of two important steps: observation and the interpretation of an animal's behavior. Interpretation or explanation in turn has its functional, causal, and phylogenetic aspects, which deal with the adaptiveness of behavior, and the underlying mechanisms and probable course of development of the behavior during its evolution. The "goal" of ethological research has been reached when it is possible to predict the future behavior of an animal or a sequence of behavior patterns and to justify the explanations.

The starting point and basis of the scientific study of behavior is a precise and detailed compilation of behavior patterns that are typical for a species that is as complete as possible. In the early years of ethology this "ethogram" consisted primarily of the analysis of protocols or notes taken during observations. Today, various aids are available that not only simplify the taking of notes but make possible a more accurate analysis and a more permanent storage of information. It is now possible to use event recorders, and especially tape recorders, which allow dictation of the observations without the need to take one's eyes off the animal. Activity can be registered automatically with photo cells, where an animal interrupts a light beam or activates a magnetic contact that triggers counters or event recorders. Videotapes, film, and audiotapes permit the analysis of movement and sound patterns in great detail. These can then be compared and quantitatively analyzed. This is facilitated by the use of slow or fast motion and single-frame analysis, and in acoustic research by the use of sound spectograms. Finally, computers make possible the rapid analysis and plotting of results.

The part of ethology that deals with the compilation of ethograms is sometimes referred to as the morphology of behavior because behavior patterns are just as unique to a species as are their morphological characteristics (see Section 3.9.1), and ethologists often use the methods of morphologists when dealing with phylogenetic questions.

From the very beginning ethologists tried not to look at the behavior of a species in isolation, but they compared it to other species as well. Good examples are the studies of Charles Otis Whitman on doves and Oskar Heinroth on ducks and geese. Such compilations of behavior repertoires of an entire genus or family, called a "group ethogram," can give us indications about the relationships within systematic groups and be a valuable contribution to the systematic classification of a species. It can further clarify the phylogenetic development of single behavior patterns (see Section 10.3). The emphasis on comparative, phylogenetic questions in ethology is reflected in the term COMPARATIVE ETHOLOGY (German: *vergleichende Verhaltensforschung*).

Another important task of descriptive ethology is to bring some order into the numerous observations, i.e., to name behavior patterns, to categorize them, and if warranted, to break them down into more basic, recognizable units. We already referred to the difficulty in finding truly objective terms that are free from the excess meaning that is present in the popular use of words. In naming behavior patterns it is best to describe their form. *Spreading of gill covers, head-nodding,* and *raising of wing* are neutral terms free of excess meaning. It is more difficult to name a behavior according to its function, at least when this occurs before the function or meaning of the behavior is precisely known. An example is the term *courtship flight* for the flight found in several species during which birds sing. The term is misleading inasmuch as most of these songs are not components of early courtship but rather they attract a female or indicate the possession of a territory. Still, the term is widely used. However, one is not justified to name a behavior pattern according to function when the behavior occurs in more than one context and hence may have several functions.

A classification of behavior can be made according to several criteria. Most important are FUNCTION and LEVEL OF INTEGRATION. In the first, a distinction is made between several "functional systems," which refer to a group of behavior patterns with the same or a similar purpose or effect. Examples are locomotion, eating, courtship, care of young, or aggression. Within each functional system further subdivisions can be made. Thus, care of young includes behavior patterns of nest building, feeding, or defense of young, while eating includes behavior patterns involved in obtaining, preparing, or in some cases the storing of food.

Another principle of classification is based on the level of integration of behavior patterns. The behavior is arranged hierarchically (see Chapter 4) from the simplest behavior elements (e.g., movement of a single muscle or group of muscles) to units of an intermediate level of complexity (e.g., the movement of individual parts of the body) up to the complicated behavior sequences that consist of many components. The less complex a behavior component, the more frequently it generally appears not only as an element of a single functional system but also in several behavior categories.

In ethological investigations it is important that the appropriate level of integration be selected for a particular problem. If one studies the dependence of song on the male sex hormone in a species, then it is sufficient to look at the entire song as a unit. However, if one is interested in the information content of a song for certain conspecifics, e.g., rivals or sexual partners, then each discriminable song component must be examined singly or in various combinations with the others.

Descriptive ethology is not limited to a mere inventory of behavior patterns of a species. Ethologists can also provide important conclusions about the organization of behavior. This requires that behavior be not only qualitatively described but quantitatively assessed as well so as to determine which behavior components occur together, which exclude one another, or which occur in what particular temporal sequence. A temporal order can be the expression of common or opposing internal causes with respect to the behavior patterns in question (see Chapter 4).

1.3. Experimental Ethology

A sequence and temporal organization of behavior patterns can be determined simply by description. Broader statements about causation can usually be made only after some artificial manipulation of the behavior has occurred. In the beginning such experiments were quite simple. This is illustrated by one of the classical examples in ethology. The female of the sand wasp *Ammophila campestris*, digger wasp, deposits its eggs into nests that she has dug in the ground and into which she carries food she has caught and immobilized or killed—usually some caterpillar. After laying an egg, she closes the nest but returns when the egg has hatched to provide the larva with food as needed. Finally, she brings three to seven caterpillars, sufficient nourishment until the larva metamorphoses, and closes the nest. Observation alone is sufficient to learn about the sequence of this broad care behavior, which consists of an orderly sequence of many components. Furthermore, one can recognize that the female's behavior is appropriate to the situation, i.e., that the size and number of the prey is adjusted to the size and hence the food requirements of the larva. It is not known how the sand wasp, which simultaneously cares for several nests that contain larvae of various ages, remembers their size and hence the amount of prey needed at each nest. This cannot be determined by mere observation.

However, even a simple experiment can help here. The female can be deceived by the addition or removal of larvae. As a result of such manipulations, the wasp will adapt her behavior to the new situation. She will bring either more or less prey than she would have otherwise. Hence, it is not the memory of her own previous brood care behavior at the same nest that determines how much food is brought, but the information available from the content at each nest. However, this adaptability of the wasp shows a peculiar temporal limitation: an artificial change in amount of prey in the nest is compensated for only when this is done prior to the early morning "inspection

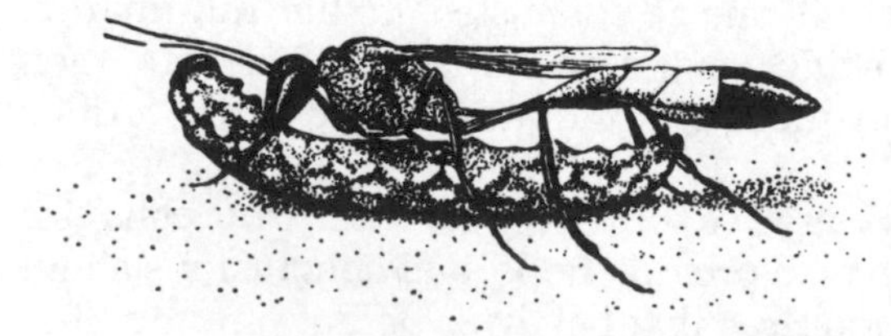

Fig. 1. A sand wasp carries a caterpillar she has caught to the nest (after Baerends 1941, from Tinbergen 1951).

visit" of the wasp. Once the female has inventoried the food situation at a nest, the information obtained then determines how much food is brought for the remainder of the day. After that, the wasp will not respond to changes in the amount of prey in the nest.

The conclusions drawn from these simple manipulations give an indication of the opportunities in experimental ethology, which employs increasingly sophisticated methods. Models play an important role as well. The word MODEL is used in ethology in a different sense from that in everyday language, where one may think of as realistic an imitation of an object as is possible. Ethologists often use quite unnatural models. Any representations at all are considered models, as long as they are effective in releasing a response in an animal so that the stimulus properties that elicit a given behavior may be analyzed. Hence, models of only part of an animal's body are used, such as a beak or head. Various properties of the object to be imitated are changed (enlarged, made smaller, presented in a different form or color). Even completely unnatural objects (e.g., wooden spheres or cubes of various colors) may be used as models, as well as tape recordings or vocalizations and artificial odors. Elements of the natural example or cues can be changed, which makes it possible to test the effective components of the stimuli. Thus, it is possible in acoustic research to detect by changing tape-recorded sounds (e.g., transpositions of components in total sequences, changes in rhythms, elimination of certain frequencies) which characteristics of biologically significant sounds are important, for example, in the song of a rival.

One of the earliest experimental investigations, in which models were used to a greater extent, dealt with the pecking reaction of young herring gulls. Hungry gull chicks peck at a red spot on the yellow bill of the adult bird and thus trigger the regurgitation of food. Which cues on the gull's head were effective in releasing the pecking response in the chick was determined with the aid of cardboard models on which various combinations of colors of the spot as well as the bill and head were painted. The results of the experiments show that the necessary cues are located on the bill, while the

Fig. 2. A herring gull chick pecks at the bill of the adult bird (after Tinbergen 1951).

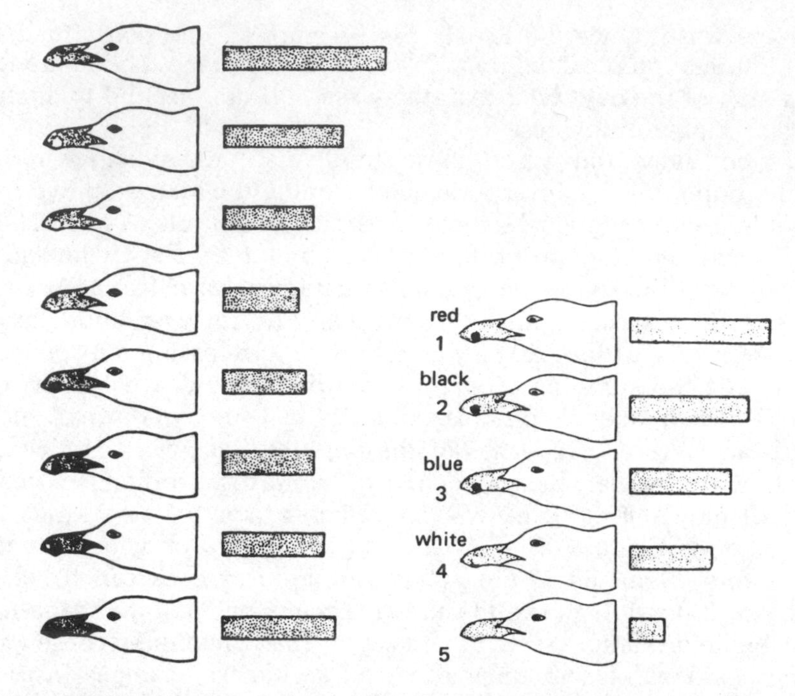

Fig. 3 (left). Cardboard models used to test the effectiveness of the spot on the lower part of the bill in releasing pecking. The length of the bars indicates the relative frequency with which inexperienced herring gull chicks peck at the model. They show that the frequency of pecking increases with greater contrast of the spot against the background (after Tinbergen 1951).

Fig. 4 (right). The color of the spot is also important: a bill with a red spot releases more pecking than one with other colors, even though they may show greater contrast to the background (after Tinbergen 1951).

shape of the head, its size, and its color have no effect. On the bill, the red spot plays an important role. It is effective by its color as well as by the degree of contrast from the background of the bill. Thus, if models with medium-gray bills and spots ranging from white to black through various shades of gray are presented, the chick's pecking is released more frequently the "whiter" or "blacker" the spot is, i.e., the more it contrasts with its background. On the other hand, a model of the natural yellow bill color and a red spot is more effective than a black spot, although the latter offers more contrast.

Similar analyses of relevant stimulus situations using models have since been carried out on a number of animal species. Usually, the models are presented either successively or simultaneously with two or more stimuli. Simultaneous presentation offers the advantage of immediate comparability for the animal, while during successive tests habituation and fatigue can lead to a situation in which the animal's state at the beginning of the test is not comparable to subsequent presentations of the stimuli. Hence, the reactions would not accurately reflect the "value" of the model.

Fig. 5. A male zebra finch courts a female model, which consists of a stuffed skin attached to a perch with a wire. This test allows one to find out if the visual, nonmoving characteristics of the female are sufficient to release sexual behavior in the male and which courtship patterns of the male can be elicited without the auditory (voice) and dynamic-visual (movement) cues normally provided by the female. In successive presentations this test can be used to examine sexual preferences. This can then be compared with tests of simultaneous presentations (see Figure 46, Section 7.4.7.1).

Various subareas of ethology have their own specific methods of testing and will be discussed in the appropriate places.

1.4. Areas of Ethological Research

It is possible to define several areas of ethology beyond the classification as descriptive and experimental ethology. These areas focus on specific questions that are treated either descriptively or experimentally. There is much overlap with other areas of biology such as ecology, physiology, and genetics. While their methods may be used and similar questions asked, these refer not to body structures but to the behavior of the animals. Sometimes it is not possible to distinguish whether a study is more ethological, ecological, or physiological. Hence, it is not always possible to delimit an area of ethology precisely.

The area of ECOETHOLOGY is a comparatively new branch of ethology in which the relationships between the behavior of a species and other living and nonliving components of environment are investigated. Ecoethology can proceed in one of two ways: it can emphasize or focus either on a group of species or on a particular habitat or biotope. In focusing on the biotope, one would be interested in the parallel behavioral adaptations that are found in certain habitats, for example, deserts or tropical rain forests. This is of interest even in species that are not closely related, and whose behavior may be considered typical for the particular biotope. In the second case, we look at a group of closely related species with an interest in how the various species differ from one another and whether or not and how such differences can be considered adaptations to various habitats, i.e., what is the "biological" significance of a behavior. Especially interesting results can be expected when within a related group of species there is one that lives in an entirely different habitat and whose behavior deviates substantially from the behavior typical for the group.

Unlike any other area of ethology, ecoethology depends on a careful balance of investigations in captivity and in the natural environment of the particular species.

An important subarea of ecoethology is SOCIOECOLOGY or SOCIOBIOLOGY. Sociobiology deals with the relationships between the environment of a species and its social structure, i.e., the organization of spatial distribution and social organization of the members of a species, e.g., whether they are solitary or live in pairs or groups. Animal sociology (German: *Tiersoziologie*) is not identical with sociobiology, even though it also deals with the social behavior of animals. However, here the emphasis is on mechanisms, i.e., means of communication that help to establish and maintain the social structure.

BEHAVIORAL PHYSIOLOGY or ETHOPHYSIOLOGY deals with the physiological basis of behavior. Two of its main branches are concerned with the two large control systems of an organism that are also important in the area of behavior. NEUROETHOLOGY deals with the sensory processes and the central nervous system that underlie a particular behavior. ETHOENDOCRINOLOGY[1] deals with the reciprocal relations between hormones and behavior.

BEHAVIOR GENETICS, sometimes called ETHOGENETICS, investigates the genetic basis of behavior with the methods of genetics. Its goal is the deciphering of the relationship between genetic factors and how they influence behavior.

Two areas of ethology deal with the change of behavior over time. The PHYLOGENY OF BEHAVIOR traces the evolutionary origin and development of behavioral characteristics, and the ONTOGENY OF BEHAVIOR studies the development of behavior in an individual. A subarea of this latter branch is BEHAVIORAL EMBRYOLOGY, which is concerned with the prenatal development of behavior patterns.

Finally, the youngest area is HUMAN ETHOLOGY, whose goal it is to study human behavior with ethological methods. It emphasizes phylogenetically transmitted and genetically determined regularities and variability of behavior.

In addition to these areas, there are several biological disciplines that do not really belong with ethology, but because they also deal with morphological and physiological questions they overlap widely with ethology. First among these is sociobiology, a very young area, which has already had a substantial influence on contemporary biological thought. With respect to subject matter, sociobiology stands between ethology and population biology, a research area that deals with the temporal and spatial distribution of individuals and their relationships to the living and nonliving environment. Like ethology, sociobiology is primarily a comparative science. It examines the biological bases of

[1] ENDOCRINOLOGY is an area of zoology that deals with the investigation of hormones, their chemical structure, their properties, manner of acting, sites of production in the endocrine glands.

various aspects of social behavior and deals primarily with the selective advantages of specific social structures (see Sections 8.4.2–8.4.6.). In contrast to ethology, sociobiology is less interested in the mechanisms that determine the social organization of a species or population, or how it is maintained.

Sociobiology has caused ethologists to pay more attention to the intraspecific variation of behavior and to individual behavior patterns. Beyond that it has offered biological explanations for those behavioral characteristics that had not previously been explained by the theory of natural selection (see Sections 10.1 and 10.2), such as the existence of altruistic behavior (see Section 8.4.3).

Two other fields that border ethology are bioacoustics and the study of biological rhythms.

The goal of BIOACOUSTICS is the investigation of vocalizations of animals. This research area has come into its own during the last two decades with the appearance of high-quality recording equipment and tape recorders. The ability to produce sound spectrograms resulted in an objective and reproducible presentation of animal sounds which supplanted the old and subjective description by words and musical notes (see Section 3.5). More generally, bioacoustics includes the investigation of auditory sense organs and the sound-producing apparatus (vocal cords, stridulating organs, feathers modified to produce sounds), as well as the study of physiological processes that are involved in the production and perception of sounds. Finally, it includes the relationships between the vocalizations of the animal and its habitat. Bioacoustics has contributed important knowledge to ethology in the area of the ontogeny of behavior and with respect to acoustic communication, i.e., communication between conspecifics and members of other species.

The study of BIOLOGICAL RHYTHMS (e.g., diurnal, monthly, and annual rhythms) is concerned with recurring, rhythmic events of living organisms and their underlying processes. Since such periodicity is so prominently expressed in the behavior of an animal, there is a close connection between ethology and the study of biorhythms.

2

Basic Ethological Concepts

2.1. Reflexes

For a long time, an animal's behavior was seen exclusively as a reaction, i.e., as a response to external or internal sensory stimulation. This was especially true for physiologically oriented investigators. The simplest form of reactive behavior is the REFLEX. It is characterized by an especially strong stimulus–response connection. Under the same conditions and in response to the same stimulus the response always occurs in exactly the same manner.

This response invariability is the result of the expression of very rigid anatomical connections between nerves that make up a reflex and that are responsible for the conduction of the appropriate nerve impulses. These connections are known as a REFLEX ARC. A reflex arc begins at a receptor, i.e., a sensory nerve cell or a sense organ. Then it passes along a so-called sensory or afferent pathway to the central nervous system, and on from there via a motor or efferent pathway to the effector organ, which is a part of this reflex arc.[1] The simplest reflex arcs consist of only one afferent and one efferent nerve cell. Hence, they share only a single synapse (MONOSYNAPTIC REFLEX). In most cases, however, a large number of nerve cells and synapses are involved (POLYSYNAPTIC REFLEX). In vertebrates many reflex arcs terminate at the spinal cord, and only a few reach the brain. Here they pass into the areas with the simplest structures in the brain stem. Hence, reflexes are "involuntary" in the sense that they are not under the control of the higher centers of the brain. Even in man they occur unconsciously.

The reaction that constitutes a specific reflex, e.g., the form of the accompanying motor patterns, is always innate. (For a discussion of the use of the term *innate* in ethology, see Section 6.1.1.) In regard to the connection between the releasing stimulus and the reaction, there are, however, basic differences. As a rule the form of the reaction in a reflex also has an innate basis, i.e., it is independent of previous experience with the specific stimulus. This kind of reflex is called an UNCONDITIONED REFLEX. It is distinguished from the CONDITIONED REFLEX, where knowledge of the releasing stimulus is

[1] An EFFECTOR ORGAN is any one of a number of organs or parts of an organ of the body that carry out some function, i.e., all structures that carry out some activity, such as movement or the secretion of a substance in response to a nerve impulse. This includes muscle and gland cells and those pigment cells that contain nerve endings, i.e., are controlled by nerves and are thus able to bring about a rapid change in color.

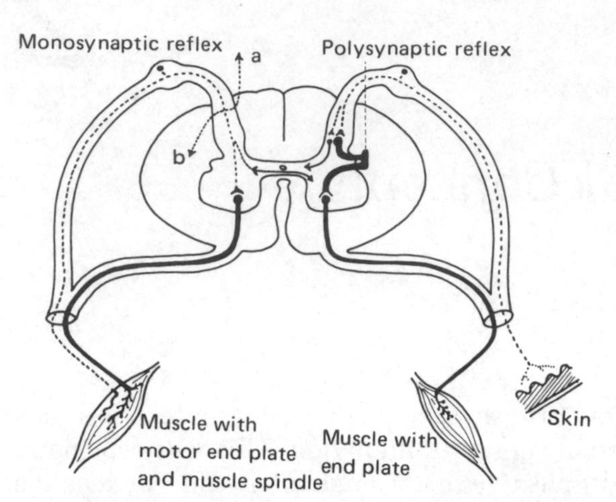

Fig. 6. Schema of a reflex arc: the sensory input arrives from a receptor (here a muscle spindle or a free nerve ending in the skin) and enters the spinal cord via the dorsal root. From the synapse originates the ventral root, which leads to the effector organ via the motor neuron. The nerve fiber divides before the synapse into an ascending branch and a descending branch (a,b), which lead to adjacent spinal cord segments. In addition, lateral branches and more synapses form cross-connections to the opposite side of the spinal cord (after Gardner, redrawn from Romer 1959).

acquired. In other words, the stimulus–response connection is the result of a learning process.

The unconditioned reflexes comprise many of the protective reflexes (scratching, wiping, withdrawal reflexes; closing of the eyelid and pupil reflexes; sneezing, coughing, and vomiting reflexes) as well as reflexes that maintain the constant length of muscles (knee-tendon reflex) and control of balance and posture. One automatic reflex that plays an important role in the history of ethology is the salivary reflex. Most reflexes are extremely short, and longer reactions are very rare. An example is the grasping reflex with which a young primate holds onto its mother (see Section 8.3.1).

Conditioned reflexes were mainly investigated by the Russian physiologist I. Pavlov. He carried out his work on the salivary reflex in dogs. This reaction is released as an unconditioned reflex by food stimuli (odors, visual cues). If, in an experiment, the presentation of food is paired with an unrelated stimulus—for example, a flash of light or the sound of a bell—then the secretion of saliva can be released after several presentations of such initially neutral stimuli even in the absence of the food. The similarity to the unconditioned reflex is that the stimulus–response relationship is also very rigid, i.e., the animal responds with great reliability to the appropriate stimulus. The external stimulus that initially triggers the reaction in an unconditioned reflex is called the UNCONDITIONED STIMULUS. The initially neutral stimulus that can secondarily release the reaction is the CONDITIONED STIMULUS.

The studies of Pavlov and his followers have made substantial contributions to our understanding of the nature of learning processes (see also Section 7.4.2). However, they also led to the rapid development of a "reflexology," which was largely responsible for the fact that the behavior of animals was long seen as based exclusively on the reflex activity of the central nervous system. It was recognized, however, that many behavior patterns must consist of long and complex chains of reflexes. Not until the middle 1930s did Erich

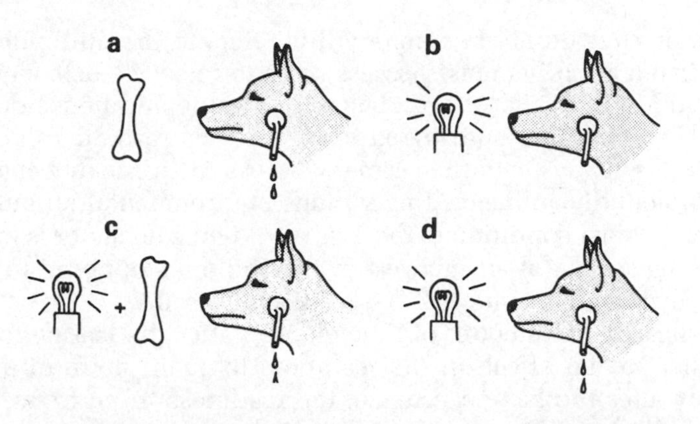

Fig. 7. Pavlov's experiments demonstrating the salivary reflex in the domestic dog: seeing the food (UNCONDITIONED STIMULUS) releases the secretion of saliva. The saliva is collected from a duct implanted in a salivary gland below the ear, and the amount of salivate is measured. This is the UNCONDITIONED REFLEX (a). A neutral stimulus (signal) releases no reaction (b). When the unconditioned stimulus and the signal are presented together (c), a salivary secretion will be released after several presentations, even when the signal is presented alone (d). A CONDITIONED REFLEX has been established (after Pavlov, from Rensing *et al.* 1975).

von Holst and Konrad Lorenz independently discover on the basis of experiments and observations that behavior not only is reactive to stimuli but is also based on internal, spontaneous causes. This discovery made it possible to separate innate behavior patterns conceptually from reflexes and to define innate behavior by the presence of such spontaneous components.

In the following chapters we shall first examine the facts that argue against the chain-reflex hypothesis of innate behavior patterns. This will also help to explain some of the regularities of the behavior of animals and will allow us to introduce some important concepts. Subsequently we shall outline the characteristics of the instinctive behavior system.[2]

2.2. Threshold Changes

Like simple reflexes, complex behavior patterns can also be released by external stimuli (see Chapter 3). However, in spite of these and some other similarities there is one basic difference: in a reflex the response to a stimulus is by and large identical from one presentation to the next, whereas in an innate behavior pattern it can definitely vary.

This difference can best be illustrated with the concept of MINIMAL

[2] The concepts of *instinct* and *drive* are only rarely used in ethology today for various reasons. They are being replaced by more neutral terms (see Section 4.3).

THRESHOLD OR THRESHOLD. In ethology this refers to the minimum amplitude or quality that a stimulus must possess so as to release a behavioral response in an animal.[3] Stimuli that remain below this value and hence do not release a noticeable reaction are considered to be below threshold value. In a reflex the threshold is always about the same, whereas in an innate behavior pattern it can be profoundly influenced by various environmental stimuli as well as by the physiological condition of the animal. When a behavior is more difficult to release, one speaks of an increase in threshold, as opposed to lowering of threshold when the behavior is more readily released.

It is primarily the amount of time elapsed after the last performance of a behavior that has an effect on the readiness to again perform the behavior. Immediately after the last occurrence, the readiness to perform the behavior again is usually low. Thus, an animal that has just copulated generally does not respond to sexual stimuli, or it does so only weakly. Similar increases in threshold exist in searching for food and in other functional systems (see Section 2.3). On the other hand, a behavior pattern that has not occurred for some time can be quite readily released, and it can be elicited by increasingly simpler and less specific stimuli. A well-known example is that of a dog shaking a slipper: members of the dog family, predators, use the behavior to break a prey's neck, to disorient its sense of equilibrium, or to keep it off balance, which all results in temporary immobility of the prey. Dogs, unable to carry out this behavior on natural prey, do so on nonliving objects. Cats carry out certain prey-catching behavior patterns on substitutes like balls and other small objects that are readily moved. In pet, zoo, or laboratory animals that live without conspecifics, threshold decreases occur especially with respect to sexual behavior. These animals direct copulatory and other sexual behavior to members of other species or even to rather crude models. Under natural conditions this would almost never occur. Such inappropriate models that trigger a normal behavior are called SUBSTITUTE OBJECTS. In extreme cases a lowering of threshold can lead to the spontaneous occurrence of the behavior, i.e., it is performed in the complete absence of any external stimuli. This is called VACUUM BEHAVIOR. It presents a problem methodologically in that it is very difficult to prove that no environmental stimuli, however inadequate, were present that may have released the behavior. Hence, well-documented examples of vacuum activities are very rare. The most impressive examples are undoubtedly the complex nest-building behavior patterns of weaver birds, which may be performed in a cage in the absence of any grass stems, their usual nesting material, and without the use of any substitutes.

[3] In sensory physiology the concept of *threshold* is also used. However, there it is applied primarily to the capacity of sense organs, while the minimal threshold, as ethologists use it, is probably a property of the central nervous system. In sensory physiology the threshold is the minimal value of a stimulus that triggers a nerve impulse in a receptor. Thus, the auditory threshold is the lowest just-noticeable amplitude of a stimulus that can be heard.

2.3. Specific Fatigue

Changes in threshold indicate not only that behavior patterns are responses that remain constant when elicited by external stimuli, but that there must also be an internal component that is responsible for the differences in the readiness to respond at a particular moment. A similar conclusion can be drawn from the phenomenon of ACTION-SPECIFIC FATIGUE of behavior patterns. This term refers to two different processes, which are called action-specific and stimulus-specific fatigue.

2.3.1. Action-Specific Fatigue

The above-mentioned changes in the ease with which a particular action is elicited, and which can in principle be observed in all complex behaviors, do not occur in all instances to the same extent. There are many behavior patterns that can be released shortly after they have been performed, while for others a much longer interval is required. This fatigue, then, is for one specific action only; it is action-specific. These differences are related to the specific requirements of particular behavior patterns: while sexual or food-related behavior, especially in animals that eat large amounts of food at a time, requires that the behavior occur only periodically, escape and defensive reactions need to be continuously ready to go. It is therefore biologically meaningful that the latter show little variability in the ease with which they can be triggered.

2.3.2. Stimulus-Specific Fatigue

However, differences in the degree of fatigue affecting behavior patterns can also occur in one and the same behavior pattern. They then depend on the stimulus that releases the reaction. Thus, a behavior that can no longer be released after several presentations by a particular stimulus may again be released by another. This can be readily observed in the prey-catching response of dragonfly larvae. The movement of the lower mandible, which is modified to enable the animal to capture its prey, is normally elicited by visual stimuli (size of prey, specific movement patterns, etc.). During repeated stimulation by visual stimuli this reaction can be temporarily exhausted. However, if one now presents a tactile stimulus the response again occurs.

Similar phenomena may be demonstrated within the same sense modality, e.g., within the auditory system. Thus, the "gobbling" vocalization of the turkey, a courtship behavior, can be released artificially by the presentation of certain pure tones. With repeated presentations of the same tone the response eventually stops. However, the response is reactivated at once if a tone of a different frequency or amplitude is presented. This indicates that the previous lack of response depended on previous repetition of the response, rather than on the repeated presentations of the stimulus. Hence, a decrease in respon-

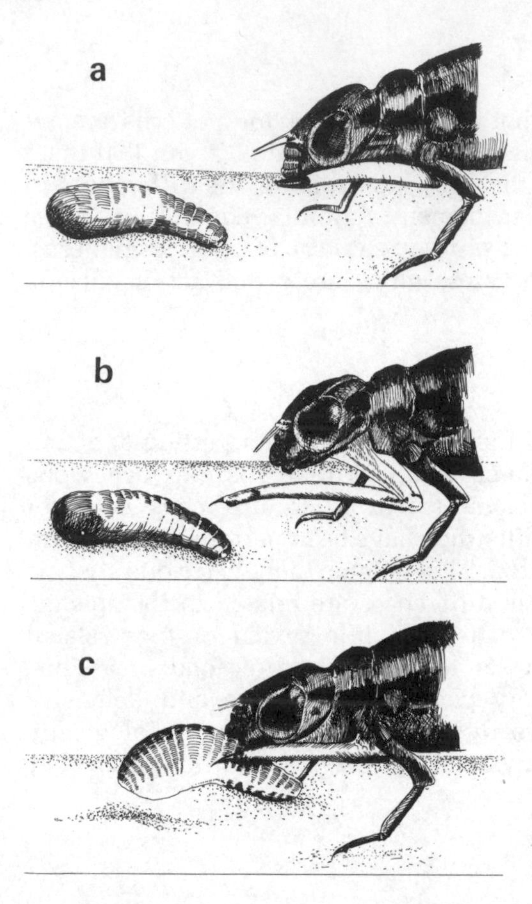

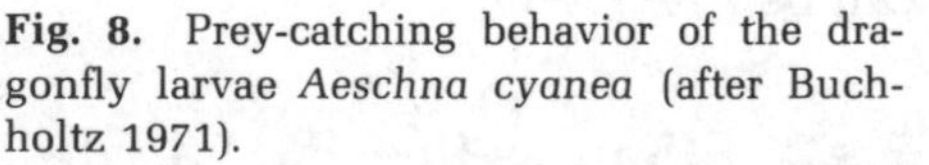
Fig. 8. Prey-catching behavior of the dragonfly larvae *Aeschna cyanea* (after Buchholtz 1971).

siveness to a particular stimulus is also habituation, or adaptation,[4] although there is frequently no agreement about these terms and the similarity or the small differences between them, as well as about the underlying phenomena. Most authors consider stimulus-specific fatigue a simple form of learning.

Both forms of specific fatigue support the conclusion that the differences in readiness to respond are due neither to "fatigue," in the sense of muscular fatigue, nor to a decrease of sensitivity in the sense organs, although we cannot exclude both of these possibilities. Instead, there seems to be a fatigue of those components of the central nervous system that pass on the afferent nerve impulses to those parts of the brain that are involved in the behavior. Since centripetal, afferent pathways are involved (see Section 2.1), one speaks here of AFFERENT THROTTLING. A sharp delineation from similar processes in the sensory and motor areas is, however, not always possible. E. Curio

[4] The term ADAPTATION is also used in other areas of biology, e.g., sensory physiology and evolutionary biology. In each case it has its own specific meaning. Its use in certain borderline cases can easily lead to confusion.

demonstrated the "central" nature of stimulus habituation in a field experiment. He studied the mobbing reaction of nesting pied flycatchers, which, like many other smaller birds, react to many resting aerial predators, especially owls, with an alarm reaction in which they utter specific calls and make twitching wing and tail movements. This behavior can also be released by appropriate models of the predator. If a stuffed pygmy owl is placed near the nest cavity of the birds, they show a strong alarm reaction. However, a wooden model painted to resemble the owl will never be mobbed. Apparently the feathers are an important stimulus for the release of the reaction. This mobbing reaction, like similar responses, is subject to fatigue, i.e., it weakens after repeated presentations of the model until it ceases altogether. A surprising and important result with respect to the problem of stimulus-specific fatigue is obtained if one first presents the ineffective wooden model and then the stuffed owl: the latter will no longer release the mobbing reaction, indicating that the readiness to respond to the normally effective model is gone. This shows that the wooden model, while unable to release the reaction, nevertheless causes stimulus-specific fatigue, and that this can occur even when the response itself has never been performed. The contrast to motor fatigue is here especially obvious.

2.4. Appetitive Behavior and Consummatory Acts

Complex behavior patterns can as a rule show two components that are clearly distinguishable. With reference to the phenomena of threshold changes

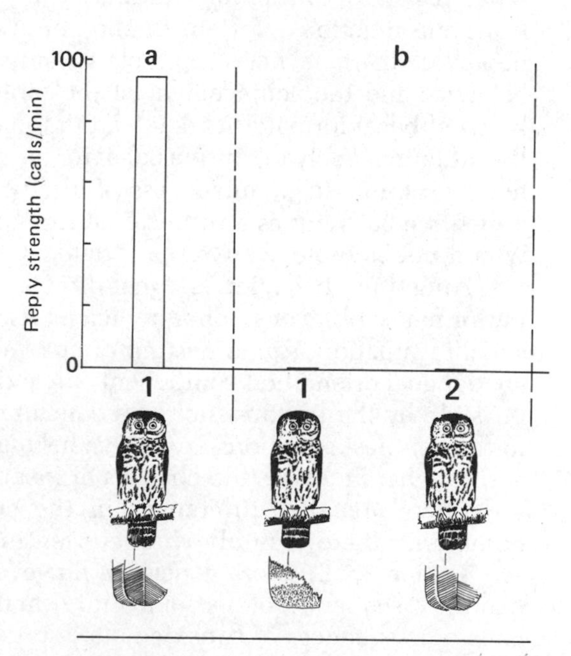

Fig. 9. The alarm reaction of the pied flycatcher, which is raising young to models of owls: a feathered, stuffed pigmy owl is a strong releaser (a), another realistically painted model without feathers releases hardly any alarm reaction (b1). After previous presentations with the model, the feathered, mounted owl is also ineffective in releasing the reaction (b2) (after Curio 1968).

and stimulus-specific fatigue they possess very different properties. One is a relatively simple and usually very stereotyped CONSUMMATORY ACTION that completes a particular behavior sequence, while the other is made up of a longer, more variable sequence of movements and orienting responses that lead to the consummatory action. This second component is called APPETITIVE or searching behavior. The distinction between them is especially obvious with respect to sexual behavior and in the functional system of searching for food. Consummatory behavior patterns consist of specific copulatory patterns and the movements involved in tearing apart and eating the food. These behavior patterns may be preceded by long sequences of appetitive behavior resulting in finding and catching the prey, or by those leading to a synchronization of sexual behavior between partners (see Section 8.2.2.4).

Consummatory actions are drive-reducing, i.e., the result of performing the behavior is that it will not occur for some time. This includes the appropriate appetitive behavior. In other words, an increase in threshold occurs which means that the same stimuli that earlier released appetitive and consummatory behavior are temporarily less effective. In contrast to this, the performance of appetitive behavior has no effect on the subsequent readiness to respond. Hence, "appetitive" behavior is not drive-reducing. It is free of action-specific fatigue (see Section 2.3.1) and can be repeatedly released.

These differences in the properties of appetitive and consummatory behavior can be explained by their different functions. With the occurrence of the consummatory behavior, the "biological goal" of a behavior sequence—e.g., satisfying hunger or fertilizing the ovum—has been achieved. The next time it would be appropriate to perform the behavior again depends on the biological necessities, and this may differ from one species to the next and from one functional system to another. With respect to appetitive behavior, however, there is no direct relationship between the performance of the behavior and the achievement of the biological goal, since the number and kinds of behavior patterns that lead to the consummatory behavior depend on the situation. Each appetitive behavior is not invariably followed by a consummatory action (e.g., unsuccessful prey-catching behavior). A decrease of appetitive behavior as a result of performance in response to the same stimuli would not be biologically appropriate.

Appetitive behavior is "goal-directed" in the sense that it leads to the performance of a consummatory act as its goal. Sometimes merely reaching a specific situation, e.g., a nesting or spawning site, is the goal. We also find a spatial goal orientation component, since the behavior that is released or made possible by the performance of a consummatory action is often some object. In the simplest situations appetitive behavior is only an increase of locomotor activity that increases the chances of meeting the appropriate object.

There are also differences in the complexity of appetitive behavior. Sometimes there are only simple orienting movements (taxis components—see Section 3.9.2). More generally, however, appetitive behavior consists of a complex sequence of various individual acts. In this case there may be behavior sequences of considerable duration. These include the migrations of birds and fish returning to their spawning grounds, which may take many months. Some authors consider such migrations to be appetitive behavior in a broader sense, since it ceases with the arrival in a consummatory situation at the breeding or spawning site.

In these and other instances of complex or temporally spaced appetitive behavior it is not possible, as with some other ethological concepts, to make a clear distinction between appetitive behavior and the consummatory actions. Thus, it is possible that before reaching a goal or performing a consummatory action an animal is temporarily interrupted in the behavior sequence. This has the appearance of a consummatory action but it does not lead to a final termination of the entire sequence. Examples are the complex hunting behavior of some predators. In the prey-catching behavior of house cats, Leyhausen (1965) found that certain components (e.g., lying in ambush, creeping up, and pouncing) possess a certain autonomy. They are actively sought out like true consummatory acts, including the directing of behavior toward substitute objects (see Section 2.2). They occur even as vacuum activities, which indicates that they seem to have an appetitive behavior of their own. In a similar manner behavior can lead to a goal situation in which members of the opposite sex come together. However, this is again a new appetitive behavior in itself, which leads ultimately to copulation as the goal. Thus, a behavior may be at once appetitive and consummatory.

Consummatory actions can often be recognized in that they are made up to a large extent of innate behavior patterns (see Section 6.1.1). Innate components may also be found in appetitive behavior, but as a rule the more adaptive, learned behaviors that better fit the variable environmental conditions predominate. Sometimes insightful behavior is also shown (see Section 7.4.6).

2.5. Spontaneity of Behavior

The phenomena of changes in threshold, of specific fatigue, and of appetitive behavior discussed so far show very clearly that there are two important differences between simple reflexes and complex behavior patterns:

- Reflexes respond to the same stimulus in the same manner, while complex behavior patterns respond to the same stimulus in variable ways.
- Reflexes "wait passively," i.e., the animal as a rule does not directly create a situation in which the particular reflex can occur. On the other hand, with complex behavior patterns the performance of the consummatory act may be actively sought out by appropriate appetitive behavior.

These differences allow us to infer that with a complex behavior not only are immediately present releasing stimuli involved but, in addition, "internal" factors within the animal play a role.

The presence of such endogenous factors was first postulated by K. Lorenz in 1937. He suspected that the central nervous system itself generated impulses that "activate instinctive behavior patterns specifically and directly." Such behavior, which appears from "within," as it were, without the noticeable participation of external stimuli, or without being sustained by them, is called SPONTANEOUS BEHAVIOR. Behavior patterns that depend exclusively on a spontaneous basis are by definition vacuum activities (see Section 2.2). Appetitive behavior can also occur spontaneously in its early phases.

All other complex behavior patterns are to varying degrees made up of spontaneous as well as reactive elements. In practice, the decision as to whether or not a particular behavior is spontaneous is extremely difficult to make since some external stimuli can not be detected by a human observer, except under special conditions. Furthermore, the same unchanging external condition itself may act—after a certain time—like an external stimulus. Hence, the predictive powers of descriptive ethology are limited in this respect and certain conclusions can only be drawn on a tentative basis.

Direct evidence for a central excitatory state can be obtained experimentally. It is of historic interest that at about the same time that K. Lorenz postulated, based on his own observations, the existence of endogenous factors as influencing behavior, E. v. Holst provided independent proof of their existence in a different discipline.

Von Holst investigated the undulating swimming movements of eels. According to the then accepted interpretation based on reflex theory, these movements originate from the contractions of a muscle element, which in turn stimulate the proprioceptors[5] of the adjacent muscle element, which then contracts. It was possible to disprove this hypothesis experimentally: if one restrains the middle segments of the eel's body, the result is an interruption in the sequence of muscle contractions, which appear again after the appropriate delay in that part of the body beyond the restraint, such as a tube. The contractions of the muscles hence could not have been stimulated mechanically by the anterior but restrained segments. This can be shown still more directly when one cuts the dorsal roots of the spinal cord that provide sensory input (see Section 2.1). Such a procedure in which the sensory afferent nerves are cut is called DEAFFERENTATION, and this results in the elimination of all proprioceptive stimulation. Nevertheless, the well-coordinated swimming movements are performed normally in a deafferented eel. This must mean that these movements are based on a central nervous organization somewhere in the CNS.[6] Hence, the complete elimination of sensory input, i.e., the

[5] PROPRIOCEPTORS are internal sensory cells that report the positions and movement of parts of the body and give information about the active and passive movements of the organism. They are present especially in the musculature of vertebrates (the so-called muscle and tendon spindles) and in arthropods (the so-called stretch receptors).

[6] The CENTRAL NERVOUS SYSTEM (CNS) refers to those parts of the nervous system located within the body that contains the majority of the cell bodies (neurons) and synapses. The CNS is the seat of the highest function of an organism. It coordinates the activity of the various organs, evaluates the messages arriving from the sense organs, and thus controls the behavior of the

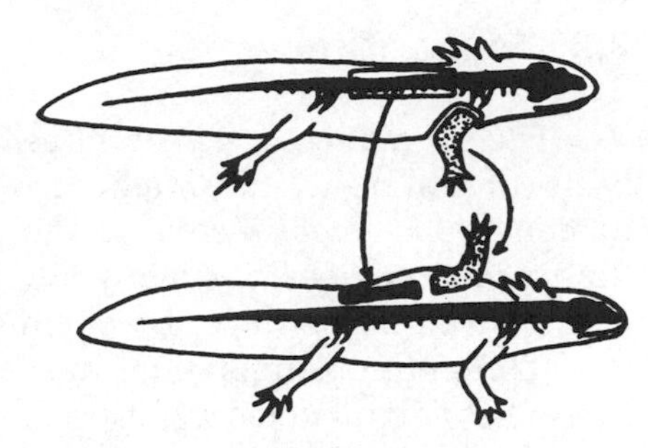

Fig. 10. Transplant of a forelimb precursor and a portion of spinal cord in the axolotl (after Weiss 1941, from Tinbergen 1952).

absence of impulses from the sensory cell and organs, does not affect the occurrence of normal behavior.

Since then, much additional support for these conclusions has been found. Deafferented tadpoles continue their specific swimming movements, and goldfish show the normal synchronization of swimming movements and breathing even in certain states of narcosis. If one transplants an embryonic forelimb bud and a segment of an embryonic spinal cord of another animal into the connective tissue of the back of an axolotl, a Mexican salamander, then nerve fibers will gradually grow from the transplanted segment of the spinal cord and enervate the limb bud. The motor nerves responsible for movement will complete the enervation slightly sooner than the sensory nerves. Hence, there is a stage of development in which the limb bud is already supplied with motor nerves, although no sensory stimuli can be received. At this stage the first movements of the limb bud can be observed. Hence, the rhythmic impulse patterns must be produced in the implanted spinal segment in the absence of external stimulation. The same observation can be made in an intact animal without an operative procedure during the course of embryonic development. Chick embryos already show movements on the third day of incubation, but respond to tactile stimuli only on day 7 or 8.

Still more powerful experimental results bearing on the question of spontaneous activity of the CNS in regard to behavior have recently been obtained from the direct measurement of central nerve impulses. They will be discussed elsewhere (see Section 5.1.1.3).

Overall, one can say that the CNS is capable of producing temporally and spatially well-coordinated movement patterns without external or proprioceptive sensory stimulus input. In the autonomic nervous system, the vegetative part involving the "automatic" regulation of heartbeat and breathing, such automatic rhythmic activity of groups of nerve cells has been known for some time. The above-cited experiments and similar ones demonstrated the same for the coordination of body movements.

---

entire organism. It is also the locus of learning and memory. The highest development of the CNS is found in vertebrates (spinal cord and brain) and in arthropods (abdominal, supra-, and subesophageal ganglia).

2.6. Motivation

The spontaneous aspects of behavior had been known or suspected on the basis of observations much earlier than they could be hypothesized or demonstrated experimentally. The precise meanings and interpretations and the various degrees of synonymity have led to many discussions that cannot be reviewed here. Since these terms contain much excess meaning and often have conflicting connotations in everyday language, they are rarely used today in the behavioral sciences, especially since they are hard to define accurately.

The term SPECIFIC ACTION POTENTIAL (SAP), also called ACTION-SPECIFIC ENERGY (ASE) or ACTION-SPECIFIC POTENTIAL, is much more precise. It refers to the fact that the endogenous components of a behavior are not general "motivators" *per se* but are specific to a particular behavior pattern. Still, this concept can result in errors because it leads to the conclusion that it is similar to the term ACTION POTENTIAL as used in neuro- and sensory physiology.

Today, the identification of factors that determine the responsiveness of an animal in a particular functional system is accomplished with more neutral terms such as TENDENCY or READINESS TO ACT, and most frequently with MOTIVATION. Motivation pertains to the readiness of an animal to perform a specific behavior. There is a certain amount of motivation for a specific behavior pattern at a given time. This value is the resultant or the sum of a number of external (exogenous) and internal (endogenous) factors, which include the following:

- ☐ *Internal sensory stimuli.* They are detected by sensory cells (so-called internal receptors) and play a role in the functional system of feeding and drinking: hunger and thirst. In other words, they refer to the need for food and water. They depend on the reports of internal receptors that are sensitive to the blood sugar level and sodium chloride content (osmotic pressure) of the blood.
- ☐ *Motivating key stimuli.* They can affect the readiness to respond with one behavior pattern released by other stimuli (see Section 3.8).
- ☐ *Hormones.* Their concentration in the blood can substantially affect the readiness to respond in several functional systems, especially in the realm of sexual behavior (see Section 5.2.2).
- ☐ *Endogenous rhythms, i.e., "internal clocks."* Certain life processes, including many behavior patterns, show regular and recurring fluctuations in the frequency and intensity that persist even in artificially created environmental conditions. Hence, they must be endogenously controlled. These effects are especially obvious in 24-hour or circadian rhythms, e.g., the general activity periods of diurnal, nocturnal, and crepuscular animals and differences in territorial songs in birds during various parts of the day. In annual variations (so-called circa-annual periodicity), we find the occurrence of migrations or of reproductive behavior or food hoarding only in certain months of the year.

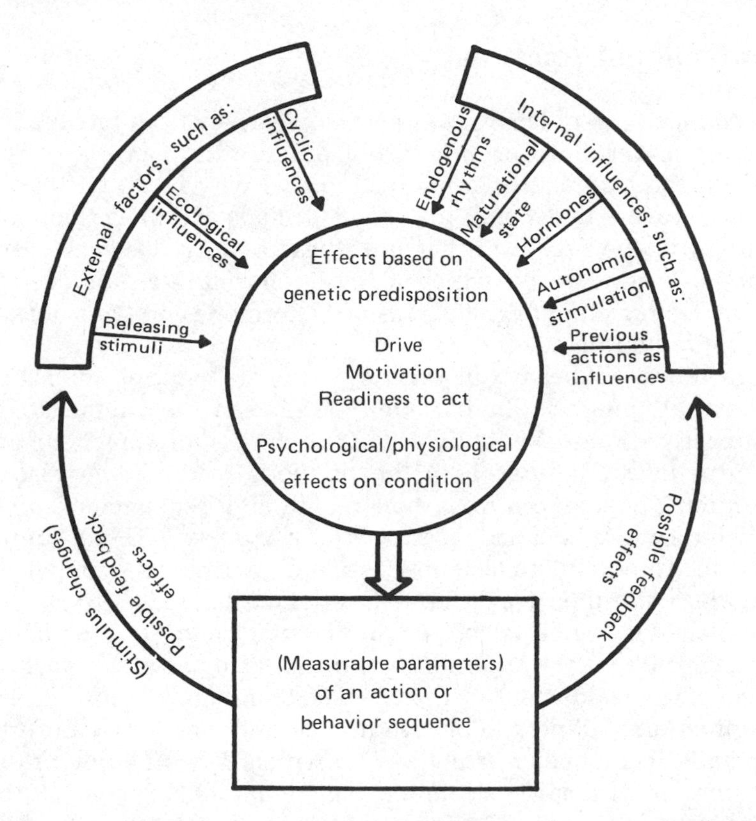

Fig. 11. Schema illustrating motivation. Shown are the measurable influences (endogenous and external factors) and the measurable actions influenced by them, which in turn affect the input (from Becker-Carus et *al.* 1972).

- ☐ *State of maturation.* The same animal may react differently to the same stimuli at different stages in its life.
- ☐ *Previous history of a behavior,* i.e., the temporal relation to the last time it was performed. Usually there is a positive correlation with the strength of the motivation and the time interval that has elapsed (see Section 2.2).
- ☐ *Autonomous production of an excitatory potential in the CNS.* It is responsible for the spontaneous behavior components, as discussed in Section 3.5.

The above-cited factors do not act in isolation but in various relationships to one another. Hence, to mention only two examples, the prior history of an action can influence the information flowing from the internal sensory cells into the CNS. Annual fluctuations of the readiness to perform a behavior are controlled by hormones and they are hence an immediate consequence of the fluctuations of the hormone level.

2.7. Motivational Analysis

This enumeration of factors may show the variety of influences that affect the readiness to act, but this should not obscure the fact that very little is known about the true nature of motivation. All we can do at this time is to observe and measure the stimuli that affect animals and their reactions. About the processes in the organism that link these overtly observable events we know only bits and pieces, which at best reflect the regular relationships between the factors cited (e.g., the effects of hormones on behavior), but they do not explain the interactions.

The readiness to behave in a certain way is also not subject to direct measurement. A quiescent animal offers no clues as to whether it is aggressively, sexually, or escape-motivated. Only the subsequent behavior can possibly give us an indication of the previous readiness to react. Hence, motivation must be deduced from the strength and frequency of the behavior itself. This procedure is called MOTIVATIONAL ANALYSIS. In the simplest case one can make important statements based on continuous or repeated observations under controlled conditions. If certain behavior patterns show fluctuations or changes under stable environmental conditions with respect to intensity and/or frequency, e.g., during the course of a day or a year or during the ontogenetic development of the animal, then conclusions can be drawn about the internal readiness to behave in a certain way. One can furthermore experimentally test whether or not and in which way an animal responds at various times to the same stimulus, e.g., a model. The most important precondition for an exact motivational analysis is that the conditions in which the animal is examined are actually the same and that learning and habituation processes or stimulus- and reaction-specific fatigue have been avoided. These conditions are difficult to achieve in practice, and the number of precise motivational analyses is very small.

2.8. Motivational Systems; "Drives"

Questions about motivation have led to countless discussions in ethology. The concept itself is not used in the same way by many investigators. Many scientists include maturational processes (see Section 6.2), learning, and fatigue phenomena in their definition because they too can affect the readiness to act, while others reject such an expanded use of the term. In addition, the general use of the term *motivation* in ethology differs from the way it is used in human psychology.

There are also difficulties in the assignment of particular behavior patterns to their underlying motivation. In most cases one can see whether a behavior is aggressively, sexually, or escape-motivated. However, there are also instances in which a complex behavior shows components of various functional systems. For example, in many threat movements and postures there are attack and escape elements (see Section 8.1.3). This is referred to as a combination of sexual and aggressive motivation. However, there are also behavior patterns occurring in several functional systems in the same form, although completely independent of other behavior components. Generally, these are relatively simple movements, e.g., pecking or snapping movements. Apparently such behavior patterns can occur in two or more motivational states.

Another question that caused much discussion concerned the number of motivation systems and the motivational independence of partial behavior sequences within larger behavior complexes. The majority of behavior patterns can be assigned to four larger motivational systems, which correspond to the "classical" major drives: feeding, reproduction, attack, and flight (escape). The latter two are not necessarily comparable with the others because they may not include spontaneous components (see Section 8.1.9.5). Then there are drives that belong to none of these four systems, such as sleep, migration, curiosity, and perhaps a separate motivation for play (see Section 7.4.4). With respect to social behavior, there are some behaviors that enhance the maintenance of a bond between pairs of a group and that may constitute a possible bonding or social drive (see Section 8.4.7).

Within the large motivation systems there are many individual behaviors—subunits of behavior, as it were—that are more or less connected, although they show considerable independence in other ways. Thus, the functional system of reproduction includes not only sexual behavior but also nest building and care of young, which in turn include collection of nest material, building, and brooding, to name a few. These separate sequences in various levels of integration, which constitute the hierarchical organization of instinctive behavior patterns (see Section 4.3), have their own drive mechanisms and may even have their own individual appetences (see Section 2.4).

Finally, we need to point out the biological significance of spontaneous drive mechanisms. Their main function is probably to "remind the animal about when a behavior should occur," i.e., when it is appropriate for the regulation of the homostatic needs of an individual, and ultimately for the preservation of a species. Hence, these spontaneous drive mechanisms insure that the first components of appetitive behavior already occur at a time when the adequate releasers for a behavior are not yet present, i.e., they lead to the search for food or a sexual partner. Only in this way is it possible to assure that the particular behavior patterns can be performed at the appropriate time.

3

External Stimuli

In the previous chapters the endogenous, spontaneous components of behavior were especially emphasized so that the differences between reflexes and innate behavior patterns would become very clear. This emphasis in presentation should not obscure the fact that innate behavior patterns are influenced in various ways by external stimuli and that, in contrast to reflexes, they are triggered, with the exception of vacuum activities, in a very different manner. Besides releasing behavior patterns, external stimuli can also influence the direction of a movement (orienting stimuli) or affect an animal's continued readiness to act (motivating stimuli).

3.1. Stimulus Filtering

Each organism is at any moment confronted by a wealth of information from the environment (e.g., about predators, food, or conspecifics), but only a small fraction is biologically significant for the animal. Hence, one of the most important tasks for the organism is the selection of those stimuli that should be followed by a reaction.

Two organ systems are available for this sorting out or filtering of information—the sense organs and the central nervous system—and both are involved in the selection of relevant stimuli. Their selective processes are called, respectively, peripheral and central stimulus filtering.

3.1.1. Peripheral Filtering

PERIPHERAL FILTERING can take place at either of two levels: either a sense organ may not be able to detect available information, or it can respond to the stimulus but be unable to pass it on to the central nervous system. The occurrence of filtering at the first level is determined by the capacity of the sense organs and differs from one species to the next. Thus, bats and some butterflies can perceive sounds in the high-frequency range, bees can see ultraviolet light, rattlesnakes react to infrared light and are sensitive to temperature differences of 0.005 degrees centrigrade, and the majority of mammals have an infinitely more sensitive sense of smell than humans. Some animals react to environmental stimuli in areas that are completely undetectable to humans and the majority of animals. Many insects are able to detect, because of the special fine structure of the visual cells in their compound eyes, the direction of waves of polarized light, an ability which enables them to orient with respect to the position of the sun even when it is not visible. Some fishes can detect electric fields surrounding them, which they themselves have produced (see Section 8.1.5.4), and the magnetic field of the earth seems to play an important part in the orientation of honeybees and some migratory birds.

The environment that can be perceived, the *Umwelt* that makes up the sum total of stimuli that have biological significance to an animal, and hence the number and kinds of external stimuli that can influence behavior, is different from one species to the next. The capacities of animals to perceive various stimuli are studied by sensory physiologists.[1]

The capacities of a species' sense organs are adapted to the requirements of its way of life. The ability to hear high frequencies is found in nocturnal moths. This enables them to detect the high-frequency sound patterns of hunting bats, and enables them to respond with escape maneuvers. The pith organs of rattlesnakes are especially sensitive to temperatures that correspond to the body heat of their prey. Generally speaking, much of the biologically irrelevant information is already excluded by the limited capacities of the sense organs of a species.

In some instances some highly specific stimulus filtering can take place. An impressive example is the female sex attractant (pheromone) found in some butterfly species that attract males from great distances (see Section 8.2.2.1). The odors are made up of chemically very complex molecules. Sensory physiological investigations have shown that the chemoreceptors, which are located on the antennae of the males, respond very selectively only to the species-specific attractant of the females and some chemically very closely related substances. They respond either little or not at all to other substances. A parallel example in the auditory modality is the males of the yellow fever mosquito, a species that lives in the subtropics. Their antennae, which resonate in response to sound waves, stimulate Johnston's organ, which responds primarily to the frequency of the female's wingbeats, while their own wingbeat frequency, which is 150 cycles per second higher, has no appreciable effect on their sense organ. Because of this, males can recognize passing females by the noise of their wings and orient toward them. Additional examples of peripheral filtering include the already-mentioned nocturnal moths, whose hearing organ responds selectively to the high-frequency sounds of the bats that hunt them. Finally, the bats often have a selective sensitivity for the frequencies of their own ultrasonic vocalizations.

All these examples have in common the fact that unnecessary information from the environment is not even perceived by the organism in the first place.

[1] In this connection, it is of historical interest that sensory physiology in its early stages used behavioral experiments extensively. Classical examples are the learning experiments by Karl von Frisch with honeybees. He determined the ability of bees to discriminate odors and colors by the use of odors emanating from glass dishes that had been placed on cardboard squares of various colors. This enabled him to determine the sensory capabilities of the bees' eyes.

However, the possibilities of such peripheral filtering are limited. Such abilities are especially useful to organisms and functional systems of animals whose need for information is very specific and specialized. This is illustrated by the example of the nocturnal moths: in many of these species adult males live only a short time after hatching from the pupae. They no longer eat: their sole function is to copulate with the females. For them there is only one biologically important stimulus: the sexual attractant of the female. A specialization of their olfactory receptors to this single stimulus is thus biologically meaningful. The same may also be true for other "specialists" such as mono- or oligophagous animal parasites, which need to respond only to one stimulus in order to find their hosts.

An exclusive or predominant filtering of stimuli by the capacities or limitations of the sense organs is possible in only a few instances, as can be seen from the examples presented here. In the majority of cases the sensory capacities are unable to achieve this kind of selection for two reasons:

(1) The nature of the biological information is too complex for an adaptation of the sense organs. Many animals react to stimulus patterns that are composed of many details and that contain reciprocal relationships that would go far beyond the capacity of a visual sense organ to filter. (2) One and the same receptor must respond to various kinds of information in more than one functional system. This need cannot be met by simple filtering at the receptor level. Hence, a stickleback male responds to the visual stimuli of daphnia by snapping, to green algae and other plants with nest-building behavior, to another male with fighting, and to a female with courtship. Even within one functional system similar problems can occur; e.g., in animals that have a varied diet many food stimuli release a positive response, while many similar but inedible objects require a negative reaction.

Hence, complex stimuli require complex filtering mechanisms. A start can be made by the sense organs in instances where they respond, because of their complexity, to an array of information, but pass on only a limited amount of the information impinging on them out of the total stimulus situation. An important role is played by so-called LATERAL INHIBITION: here a sensory cell, which has been excited by a stimulus and which passed on this excitation, can at the same time decrease or even prevent the excitation of a neighboring sensory cell. This interference results in a narrowing and enhanced contrast of a stimulus pattern. Hence, the information content per temporal unit is decreased by several orders of magnitude as the excitation from the stimulus patterns passes from the peripheral sense organs to the central nervous system.

3.1.2. Central Filtering

In spite of the many possibilities for filtering information at the periphery, these are usually not sufficient and must be augmented by the additional selection of stimuli in the central nervous system. Mechanisms and localization of central filtering, which probably play a role in the response to nearly all complex stimulus patterns, are still largely unknown. These investigations are part of neurophysiology. Ethological research can only establish or suggest the presence of the capacity for central stimulus selection. This is best accomplished in experiments involving the use of models. Here individual parts of a complex stimulus pattern can be presented separately and the necessary and superfluous components of a stimulus configuration can be determined.

Very impressive results have been obtained in experiments with the grayling butterfly. During the breeding season the males of this species perform a "pursuit flight" or "courtship flight" that entices passing females to land near them. After a short courtship they copulate with them. By the use of models that varied in color, form, brightness, and movement it was determined which stimuli from the flying females release the pursuit flight. It was discovered that the three first-mentioned cues are almost completely ineffective, but that the important stimuli are movement and contrast, i.e., stimuli emanating from a fluttering butterfly that contrasts against a background. Hence, the best stimulus was a black cardboard model dangled on a fishing rod and moved so as to rise and dip like a butterfly. If a grayling is tested only in the functional system of courtship, one gains the impression that the animals are color-blind. However, from other observations it is known that while in search of food they clearly prefer blue and yellow colors, i.e., they can clearly distinguish them. In other words, graylings are not color-blind, but during courtship they behave as if they are.

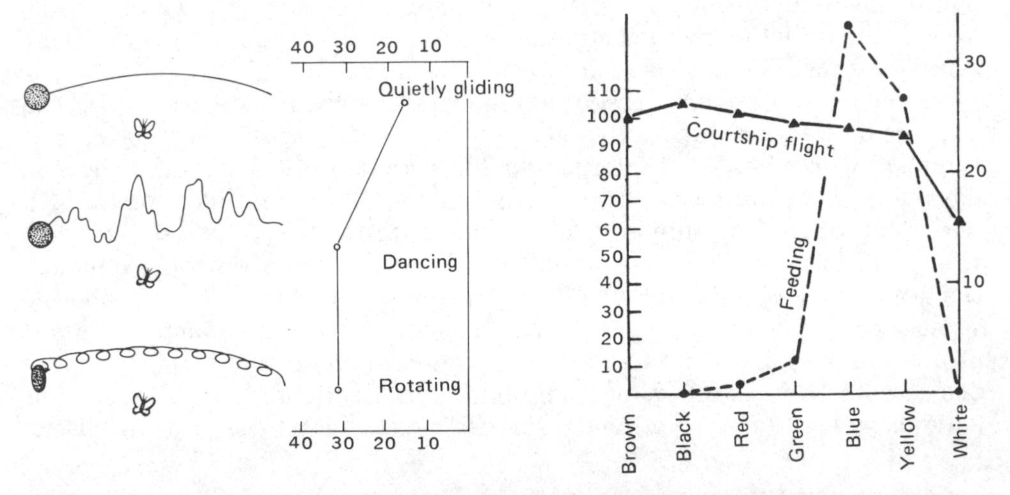

Fig. 12 (left). Experiments with models show the significance of the female's flight characteristics for the release of the courtship approach flight of the male grayling butterfly. Rotating models, which bob up and down like butterflies, are more frequently approached than smoothly gliding ones (modified after Tinbergen *et al.* 1942, in Tinbergen 1951).

Fig. 13 (right). Differential responses of color cues in two different functional systems: in male grayling butterflies color is not important during the courtship flight, but it is when visiting flowers in search of food (after Tinbergen *et al.* 1942, in Tinbergen 1951).

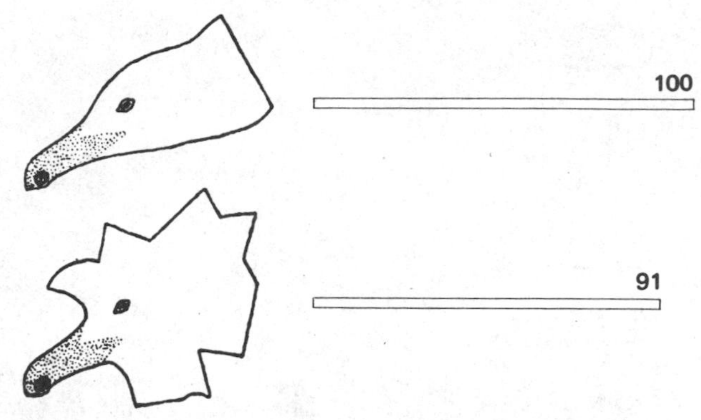

Fig. 14. The shape of the head is not important in releasing the pecking response of young herring gulls: the model with the unnatural head shape receives nearly as many responses as the natural one (after Tinbergen and Perdeck 1951).

Observations and experiments on other species indicate that the grayling butterfly is no exception in this respect: young herring gulls attend only to cues on the bill of their parents while pecking at it. Other cues such as the shape, color, and size of the head, even on quite unnatural models, release the pecking response if they possess the proper stimuli on the bill (see Section 1.3). Turkey hens that have never brooded young react with parental care only to the calls, but not to the appearance and movements of their chicks, although they can certainly perceive them visually. Even a quite unnatural model will be brooded if chick calls can be heard from a small loudspeaker mounted within. However, while turkey hens whose hearing organ has been removed will incubate their eggs, they will kill their own chicks after they hatch.

Similar limitations can also be found within one stimulus modality.[2] Thus, male sticklebacks react with aggression almost exclusively to the bright red underside of a rival, while the form, size, and surface properties are ineffective within a wide range. A natural model of a stickleback, whose underside is gray, releases less aggressive behavior than an oval wooden disk painted with a red underside. During courtship, however, the same male is quite capable of recognizing a female that is ready to lay eggs (swollen abdomen). A male European robin reacts similarly to the red breast feathers of a rival, but not to his entire shape. A red tuft of feathers mounted on a branch in the bird's territory will be threatened, while a faithful model of a robin without the red breast feathers remains unmolested. The effect of the male markings is even stronger in the Cuban finch. Not only are paper rolls and other unnatural objects threatened, but they are actually attacked if they possess a yellow mark that imitates the neckband of a male.

[2] The concept of STIMULUS MODALITY was originally used in human psychology. It refers to a certain category of stimuli that are detected by the same sense organ and that release similar sensations, e.g., light, tones, smells, or tastes. This concept was taken over by sensory physiologists when stimuli were first classified, but it is no longer in general use today.

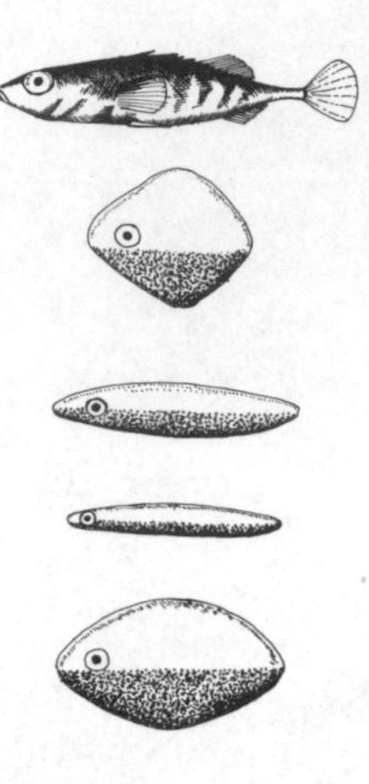

Fig. 15. Models used to test stimuli that release aggressive behavior in the male stickleback: the natural fish model in gray (top) releases fewer attacks than models of quite unnatural shapes but whose undersides are painted red (after Tinbergen 1951).

How can such a specificity or limitation of a reaction be explained? It appears that it is of greatest benefit to a species to react simply and without fail to only very few stimuli, as long as they adequately identify a biologically important object (predators, prey, nesting or hiding places). Other cues are "unnecessary" and perhaps even distracting—hence, they are best ignored. In the territory of a European robin there is no bird other than his own female that has a red breast. If one does appear, it has to be a rival. The same is true for the red underside of the male sticklebacks. Hence, it is adaptive when this cue "blindly" triggers appropriate aggresive behavior in the territory owner. Such limitations of a response are easier to develop, the fewer animals bearing a specific cue occur in the environment of a species. Conversely, it becomes more difficult to establish such responses when potential confusing species live together in one area (see Section 8.2.2.5).

Fig. 16. In the red-breasted robin, red color is a stronger releaser of aggressive behavior than the normal body: a bundle of red feathers is more intensely attacked than a stuffed bird without its red breast feathers (after Lack 1943 in Tinbergen 1951).

3.2. Releasing Mechanisms

In connection with stimulus filtering and the selective release of behavior patterns depending on the motivational state of the animal—which had already been recognized for some time—the concept of the RELEASING MECHANISM was introduced into ethology quite early. The concept was frequently controversial and is still discussed today. A releasing mechanism (RM), also called a stimulus filter or filter mechanism, is the totality of all parts of the nervous system, including the sense organs, that are involved in the filtering of all incoming stimuli. They insure that only the "appropriate" stimuli release a particular behavior pattern.[3] Appropriate in this scheme are those stimuli that characterize a biological situation in which the released behavior pattern can function effectively, for example, prey stimuli that release catching behavior, or stimuli of young that trigger parental care behavior. Thus, a releasing mechanism determines the readiness of an animal to respond to a stimulus or stimulus combination. In each functional system only those stimuli can release a behavior pattern that "fit" the appropriate releasing mechanism. Hence, the RM for visiting flowers responds to colors in the grayling butterfly, the RM for courtship flight in pursuit of the female reacts to certain movements and contrast with the background (see Section 3.1.2). These very definite statements should not, however, obscure the fact that we know very little about the nature of this neurosensory mechanism (RM). We do not know where it is located, other than that it has to be somewhere between the peripheral sense organs and the motor centers, and that it possibly involves several successive components. Hence, with respect to what actually takes place within the organism, the hypothetical construct of the releasing mechanism can contribute very little by way of explanation.

[3] The most generally accepted definition of a releasing mechanism is that of W. Schleidt: "The totality of all structures of the organism which are substantially involved in the selective release of a reaction (excluding the motor components)."

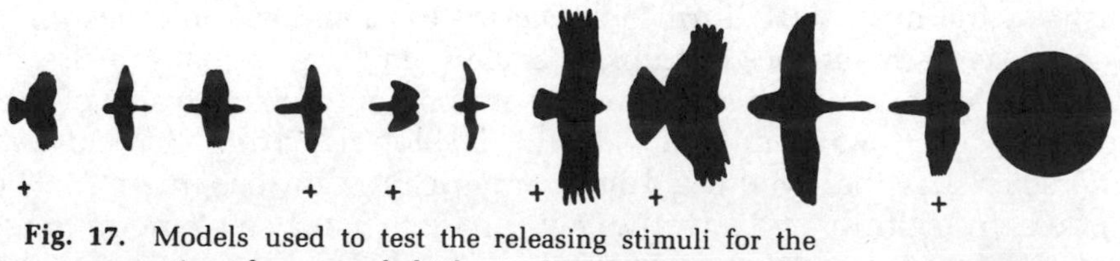

Fig. 17. Models used to test the releasing stimuli for the escape reaction of geese and chickens when they were moved across experimental birds on a wire strung overhead. Models with a + elicited alarm behavior. They all have a short neck (after Tinbergen 1951).

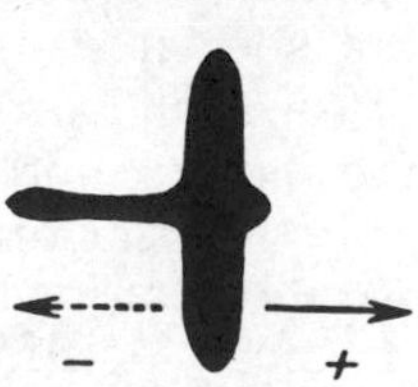

Fig. 18. The same model can either release escape or not, depending on the direction in which it is moving (after Tinbergen 1951).

3.3. Key Stimuli

Stimuli that release a reaction are called KEY STIMULI. The RM was historically seen as something like a lock that could only be opened with the right "key," consisting of the proper stimulus configuration.

Any cue that can be perceived by an organism that fulfills the criteria of the specific function listed earlier can serve as a key stimulus. Key stimuli can consist of single cues (e.g., odor or taste stimuli, tones, colors, or patterns) as well as complex relational and Gestalt cues. Hence, turkeys recognize flying raptors by their characteristic short-neck, long-tail silhouette. Young thrushes recognize the "head" on a cardboard model not by its actual size but by the relative size of the "head" to the "body."

3.4. Releasers

The concepts *key stimuli* and *releaser* are frequently used interchangeably in the ethological literature, or else no clear distinction is made. With respect to the transfer of information and the adaptation of an organism to the perception of information, there are, however, two distinct situations that justify the use of separate terms. When some particular information is useful only to the receiver, only it can evolve the optimal adaptation of the means by which to perceive this information. If an exchange of information is mutually beneficial for sender and receiver, then both can contribute to an improvement of communication through appropriate adaptations.

The first instance is primarily found in interspecific, the second in intraspecific, relations or interactions. This is best illustrated with some examples: plants, other species (prey, food competitors), and inanimate objects are recognized by an animal only by the specific cues that they already possess. However, within the same species additional structures can develop during the course of evolution that serve only intraspecific communication and hence improve it. To use an extreme example, a tree never grows colored branches merely to attract a particular bird species to nest in its branches. Instead, the bird must learn to recognize the most suitable nesting sites by such cues as the shapes of trunk, branches, and leaves that are already a part of the tree's own adaptations to its environment. In the realm of the predator–prey relations both will "avoid" developing structural and behavioral characteristics that would aid its enemy or prey in recognizing it. Predators approach stealthily and silently. Owls produce hardly a sound in flight due to the special structure of their wing feathers. Prey animals may hide and "freeze" and are thus camouflaged. Here the need for information is only with the receiver and any improvement in obtaining information can be attained only by an increased acuity of the sense organs.

In intraspecific relations, however, such as in courtship or care of young, the need for information is mutual. Both have an "interest" in good communication. The male must recognize the female and her readiness to mate. He must respond to her courtship behavior with appropriate behavior, and the female must react correctly to the male's behavior. For this reason the adaptations can in principle occur in both: the receiver can improve its sense organs, and the sender can develop stimulus patterns that are more easily recognized, which hence have greater signal value.

Releasers, then, are components of a reciprocal communication system. In their evolutionary development mutual understanding was the predominant need. The main function of releasers is the transmission of information; that of key stimuli is an accidental, additional function: the red spot on the lower mandible of a herring gull evolved exclusively as a signal for its young; the long-tailed aerial silhouette of a raptor reflects functionally determined body proportions and only "accidentally" serves as a cue in recognition by its prey.

Because of these obvious distinctions, K. Lorenz coined a separate term for key stimuli used in intraspecific communication and called them RELEASERS. He considers a releaser "any differentiation of morphological structures or innate movement coordinations whose species-preserving function is to provide signals that can be responded to in a regular manner by conspecifics" A shorter definition would be: releasers are structures or behavior patterns whose task is to release a response in one's partner.

Releasers can be visual cues (color, form), acoustic cues (vocalizations), or chemical cues (pheromones[4]) as well as behavior patterns and body postures that have evolved into frequently conspicuous and "ritualized" signals (see Section 10.6).

To fulfill their main function, communication between conspecifics, releasers must be easily recognized and conspicuous. It is not an accident that the most conspicuous structures found among animals are almost always releasers. Especially prominent examples are the tail display of a peacock (see Figure 19), the colorful plumage of male pheasants and ducks, or the luminous gape markings and mouth papillae of many young birds.

Many releasers of agonistic behavior and of sexual behavior that serve to

[4] PHEROMONES, also called ethohormones, sociohormones, or social hormones, are hormonelike substances that are produced in certain glands and, in contrast to hormones, given off externally. Their effect, then, is not within the producer but in other individuals of the same species. Pheromones have two separate functions, of which only one is directly related to behavior. Since some confusion exists in the literature, we will briefly discuss them here.

In a restricted sense, a phermone is a substance with direct physiological consequences, which, as an "externally effective" hormone, is comparable to the effects of a "true" hormone. The best known example is that of royal jelly, which is secreted by the large glands behind the upper mandibles of the queen bee and which inhibits the development of additional queens in the hive. Pheromones also play an important role in the regulation of insect castes in colonial insects.

In a wider sense, some substances referred to as pheromones have no immediately known physiological consequences but are exclusively used in social communication, i.e., they are chemical releasers. These include the sexual attractants (see Section 8.2.2.1), odor substances that mark a territory (see Section 8.1.5.4), and alarm substances that are given off by many species of schooling fish (see Section 8.4.1.1) and by some ants when there is danger.

In some instances, e.g., the royal jelly of the queen bee, the pheromone also serves as a sex attractant during the courtship flight.

Fig. 19. Some especially conspicuous releasers are often found on structures that can be folded up. They can be turned "on" and "off"—as here with the upper tail coverts of a courting peacock.

isolate species (see Section 8.2.2.5) must also possess a second characteristic: they must be highly "improbable," i.e., not easily confused with comparable markings of related species. This is especially important where closely related, sympatric species inhabit the same area and where the potential for hybridization is high (see Section 8.2.2.5).

The conspicuousness of releasers, which has developed during the course of evolution by mutation and natural selection (see Chapter 10), does have some limitations. These depend on (1) the need for camouflage and (2) the requirements imposed by other functional systems. Consequently, many releasers represent a compromise among the several selection pressures involved in their evolution.

The compromise nature of releasers is obvious when the need for camouflage is involved. Conspicuous releasers frequently develop on organs that can be folded up (eye spots on the fins of fish, colored specula on the wings of ducks, the peacock's tail feathers, inflatable throat sacks in many birds). They are usually found only—or predominantly—in males,[5] or they develop only at the time when they are actually needed. Examples are changes between conspicuous breeding or courtship plumage and cryptic camouflage coloration among fishes and birds. The intraspecific transfer of information is thus

[5] The restriction of conspicuous and hence endangering releasers to the male is biologically meaningful since in most species females are the "biologically more valuable" sex. A male can in principle fertilize the eggs of several females and hence the loss of a male is less important than that of a female.

provided by conspicuous signals when needed, yet danger to the animals is minimized at a given time or in a given individual. This phenomenon is especially prominent in some fishes—namely cichlids—which are able to change colors and hence their appearance quickly with the aid of appropriate neural mechanisms. This enables the fish to exhibit the signals only when they are needed, and furthermore signal specific moods in various functional systems.

3.5. Interspecific Releasers

Originally, the concept of a releaser was limited only to intraspecific interaction. Meanwhile, it has become known that a mutual exchange of information also occurs between members of *different* species. Such cases seem to justify the use of the term INTERSPECIFIC RELEASERS. An example is given by the alarm calls of many songbirds, which are quite similar to one another, in contrast to their other vocalizations. Hence, they warn not only their own conspecifics of danger but also members of other species (see Section 8.4.1.1). Some species react to alarm calls even when they do not possess one of their own.

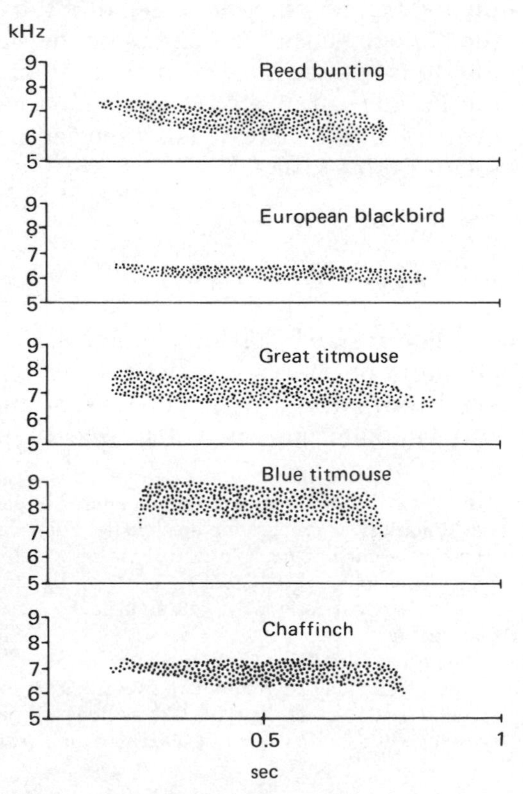

Fig. 20. Alarm calls of five species of songbirds when detecting danger from the air, e.g., a passing raptor. The calls are quite similar in frequency (between 6000 and 9000 cycles per second) and duration. They are also "understood" by members of the other species, hence they are interspecific releasers.

The figure is an illustration for the presentation of animal vocalizations by means of a sound spectrogram. On the abscissa is the duration; on the ordinate, the frequency of the sounds. In the original spectrogram the degree of blackness of the recording also gives information about the loudness of the calls (after Marler 1956).

Interspecific releasers are especially widespread in two areas: first, in mimicry,[6] where the sender imitates the signal of a model, i.e., where he contributes to the development and improvement of the appropriate signal structures and hence to the improvement of transfering of information; and second, in symbiosis,[7] where certain signals contribute to mutual recognition and understanding between the symbionts. Such a communication system is found in the so-called cleaning symbiosis between fish. Several fish species specialize in feeding on the skin parasites of other fish. They are usually recognized by conspicous colors and an equally conspicuous swimming behavior. These signals inform their "customers," which include predators that treat other fish of the same size as prey, that they are cleaners. Some species even assume a special body posture, e.g., opening of the mouth and spreading the gill covers. Cleaners feeding within the mouth are given a signal, such as closing the mouth a few times, when the fish want to move on. The cleaners then leave the mouth either through the mouth or through the gill covers.

The difference between a key stimulus and a releaser, then, is not whether or not information is transmitted between conspecific and nonspecies members, but rather whether the biological benefit is one-sided or mutual. The intraspecific signals for which the concept of a releaser was originally reserved are now referred to as SOCIAL RELEASERS by some authors. An exception to this distinguishing criterion is found in the example of mimicry cited above: here the biological benefit is clearly on the side of the sender, although it shows all the signs of a reciprocal communication system. However, this is only because the imitator "parasitizes" a signal system of reciprocal communication that evolved between two other species between which a mutually beneficial relationship exists.

3.6. Stimulus Summation

For many behavior patterns there exists not only a single releasing stimulus but several, which trigger an appropriate behavior pattern either singly or in conjunction with other stimuli. In this situation the several stimuli can facilitate one another in releasing a response. This phenomenon, first

[6] MIMICRY is the imitation of another animal or parts of another animal's signals, which results in a biological advantage for the imitator. Best known is false warning coloration: a protected species, which is capable of defense or which may taste bad, is imitated by an unprotected species, such as the mimicking of wasps by other insects. Morphological characteristics (body shape and coloration) as well as behavior patterns (e.g., certain defensive postures) may be mimicked.

[7] SYMBIOSIS is the condition in which members of two different species live together for their mutual advantage. Symbiotic relationships are found among plants (e.g., mushrooms and single-celled algae, which live together as lichens), between plants and animals (e.g., pollination by insects, birds, or bats), and between animals (e.g., hermit crabs and sea anemones).

described for the fighting and threat behavior of a cichlid species by A. Seitz, was called STIMULUS SUMMATION. It can best be studied with the use of models, since this enables the experimenter to present the various stimuli individually or in various combinations. Hence, the retrieval of an egg that has rolled out of the nest (egg-rolling movement; see Section 3.9.2) depends in the herring gull on the size, coloration, and spotting of the egg. A larger egg is more effective in releasing the response than a smaller one that is the same in all other respects. A spotted egg is more effective than one of uniform color; and a green egg is better than a brown one in releasing the behavior. A model that combines all these cues is most effective. On the other hand, a missing cue, e.g., spots, can be compensated for by an enhancement of another, for instance, by an increase in the size of an egg. It was found in female ringdoves that egg laying is induced by stimuli present in a male (see Section 8.2.2.4). The fastest response is achieved by placing a male and a female together. Egg laying can also be induced by playing dove vocalizations from the bird colony into the isolation chamber containing a female. On the other hand, a deafened female can be stimulated to produce eggs in the presence of a courting male. In the first situation, only auditory stimuli are altered, while in the second, only visual and tactile stimuli are presented. Both stimulus categories are interchangeable. Similar phenomena are known in other behavior systems as well.

This phenomenon of stimulus summation is not limited to stimuli that affect different sense modalities. The same stimulus, when presented by the same sender repeatedly, or by several senders at the same time, can have an increased effect. This seems to account for the "communal courtship" in some species where several males gather and seem to have a strong attraction for the females of their species (see Section 8.2.2.1).

The total value (effect) of a stimulus situation depends on the releasing value of all the separate stimuli that can be substituted for one another within certain limits. The use of several terms like STIMULUS SUMMATION, STIMULUS SUMMATION PHENOMENON, and LAW OF HETEROGENOUS SUMMATION can sometimes lead to confusion. Quantitative investigations have shown that individual stimuli rarely combine in a simple additive fashion to produce an effect. More generally they facilitate one another mutually without an exact summation of individual stimuli in a particular situation. This might be called a weighted-additive effect. Hence, the term RECIPROCAL STIMULUS ENHANCEMENT may be preferable.

3.7. Supernormal Releasers

Sometimes the natural signals of a species are not the best releasers in the sense of information transfer, but instead they are surpassed in effectiveness by other stimulus patterns. A stimulus that releases a particular behavior

Fig. 21. An oyster catcher incubates a supernormal clutch of five eggs, which is preferred over the usual three (a). The bird prefers the giant egg over the egg of a herring gull (left) and its own (right front) when attempting to retrieve it (after Tinbergen 1948).

pattern more effectively than the appropriate natural stimulus is called a SUPERNORMAL RELEASER.[8]

The phenomenon of the SUPERNORMAL RELEASER, which is a kind of stimulus summation phenomenon, was first discovered in experiments involving the use of models. If ground-nesting birds, e.g., oyster catchers, killdeer, herring gulls, or greylag geese, are presented with eggs or egg models that are more conspicuous or larger than their own eggs, then they are retrieved into the nest or incubated in preference over the species' own eggs. Thus, herring gulls incubate artificially colored blue, yellow, or red eggs longer on the average than normally marked eggs. Oyster catchers prefer to sit on five rather than their normal three to four eggs. Grayling butterflies approach the higher-contrast black female models more frequently than they do normally colored ones (see Section 3.1.2), and silver-washed fritillary butterflies prefer models that simulate a wingbeat frequency that is much faster than the normal one, i.e., that offer a higher number of stimulus changes per unit of time. Male glowworms prefer models with a larger light surface and yellow light to their own females even when the models present light patterns that differ substantially from that of their own species (see Section 8.2.2.1). Young herring gulls peck more frequently at a red, thin, and pointed rod with three white rings than at a realistic herring gull head made of plaster of Paris. Finally, siskins react more strongly to their species-specific contact calls if certain frequency components of the call are filtered out than if they are presented with their full natural frequency spectrum.

[8] An earlier term, SUPEROPTIMAL RELEASER, is no longer used since *optimal* already means the best possible releaser, hence one that cannot be surpassed in effectiveness.

The existence of supernormal releasers shows very clearly how strong the selection pressures are to make releasers ever more conspicuous. However, in a natural situation there are also opposing pressures to this development that then result in the compromises with respect to releasers that were already discussed. Even in the examples cited, the selection pressures favoring camouflage counter a continued development toward greater conspicuousness in signal structures. This applies especially to the egg coloration of ground-nesting birds, and to the cryptic coloration of female butterflies that is adapted to the background. However, the requirements of other functional systems may also oppose a continued evolution toward greater visibility. Thus, a thin, rodlike bill would be unsuitable for feeding in the herring gull, and the wingbeat frequency of a butterfly is subject to the aerodynamic requirements of its flight, and not only to the enhancement of the signal to conspecifics.

There are some supernormal releasers that occur in the natural situation, i.e., in the case of brood parasites and social parasites. Thus, a young nestling cuckoo (*Cuculus canorus*) possesses a gape marking in his mouth that is more prominently colored than the gape marking of its host species' young. Thus, the releasing power of its gape marking is so effective that the responses of the foster parents to their young brood parasite appears—in the words of Oskar Heinroth—almost like an addiction. In the area of chemical communication there are comparable examples in the larvae of beetles that live in the nests of certain ant species. They secrete substances from their skin glands that release brood-care behavior in the ants, such as feeding and licking. These substances mimic the ants' secretions, and in some cases are even more effective, so that the parasites are even preferred objects of brood care for the ants.

3.8. Differential Effects of External Stimuli

As explained earlier (see Section 3.1.1), external stimuli can affect behavior in various ways: they can release a behavior and determine its orientation, or they can change the readiness to act in an animal, and hence affect the threshold of a behavior pattern to respond to other stimuli. Frequently, two or all three of these effects are under the control of one and the same stimulus, but there are also examples of a clear separation of components: thus, butterflies can be deceived into landing on variously colored papers by spraying them with volatile oils. The olfactory stimulus releases searching behavior while the visual stimulus of the paper color determines the direction of the approach flight. Daphnia rise to the water surface as the CO_2 content of the water increases to reach the surface, which is higher in oxygen. Here the chemical stimulus leads only to an increased rate in swimming movements, while the orientation is to the direction of the light from above. Thus, it is possible to attract the daphnia with an artificial light source to a side or the bottom of the aquarium into even less oxygenated areas. Herring gull chicks seek cover among plants when they are in danger. This behavior is released by the parents' warning call, but orientation is determined by visual cues of the hiding place. This can result in "accidents" in certain unnatural situations, as reported by N. Tinbergen: An observer sitting in a blind near a herring gull nest made a careless movement, which released an alarm call in the parent birds. The young reacted with fleeing and ran to the nearest hiding place, which happened to be the observer's tent. Here they crouched inside at the feet of the observer who was the cause of the warning call in the first place.

Many external stimuli can also have a long-term effect and affect an animal's readiness to act instead of or in addition to their releasing or directing effects. Thus, female house mice react only weakly with parental care behavior to a day-old dead baby mouse. If one offers them first a live baby, a much stronger stimulus, then the reaction to a dead baby presented later is much stronger. It is not even necessary for a female to touch or care for the baby to obtain this result. Exposure through the bars of an adjacent cage is sufficient to bring about an increasing interest in the baby that lasts for several days.

The exact nature of this stimulus effect on the specific readiness to act is not known. Learning and sensitization processes and, especially in the case of long-term effects, hormones that are released by the preceding stimuli must play a role.

Finally, it was discovered that not the stimulus as such, but a *change* or *absence*, i.e., the difference between what was "expected" and what took place, can release a behavioral reaction. Thus, the sudden absence of a song given by a territorial songbird (e.g., one temporarily removed by an experimenter) can have a greater effect on his neighbor (e.g., invasion into the area avoided until then) than the regular singing.

3.9. Dependence upon External Stimuli

Two categories of behavior can be distinguished in the behavior of an animal with respect to their dependence on external stimuli:

- ☐ Movement patterns that merely require a stimulus to set them off, after which they continue to run their course.
- ☐ Much more complex behavior sequences whose form is variable and depends continuously on additional external stimuli.

The first category of behavior patterns is called a species-specific or species-typical behavior sequence, or most frequently a fixed action pattern, while the second refers to taxis components.

3.9.1. Inherited Movement Coordinations (Fixed Action Patterns)

FIXED ACTION PATTERNS are, according to Konishi's definition, "temporal and spatially arranged sequences of muscle contractions which produce biologically appropriate movement patterns." They are released by specific external stimuli, but these are not necessary to maintain the continued discharge of the fixed action patterns. The coordination of the various components takes place in the central nervous system and does not depend upon the kind and direction of external stimuli that affect the behavior at a given time. The only way that external stimuli can affect a fixed action pattern, besides their releasing function, is with respect to their intensity and speed. A fixed action pattern can, depending on the strength of the external stimulus, run its course at various speeds and in varying intensities while largely retaining its form. As a result of this "form constancy" or "form stereotypy," fixed action patterns are typical for a particular species and can identify it as can morphological characteristics, as Oskar Heinroth had recognized. Furthermore, form constancy is an indication, although not a proof, that fixed action patterns are inborn, i.e., that they are preprogrammed as a whole in the genome (see Section 6.1.2).

Fixed action patterns are especially obvious in functional systems such as body maintenance (scratching, cleaning, and shaking), in courtship behavior, in nest building, and in food-getting.

3.9.2. Taxis Components

The orienting components of behavior patterns are quite variable in comparison to the stereotyped form of the fixed action pattern. They depend in their *form* upon external stimuli, which not only trigger them but also direct them constantly during the performance. The difference between the two behavior components is best shown by the example in which it was first described, the egg-rolling movement of the greylag goose. This is a behavior found in many ground-nesting birds by which eggs that are located just outside the next are retrieved into the next. This involves stretching of the neck, placing the chin over the egg, and pulling back the neck with lateral balancing movements that keep the egg from slipping away. If one removes the egg while the bird is still pulling it back, the animal continues the retrieving movement but omits the lateral balancing movements. Egg rolling, then, is a fixed action pattern that, once initiated, continues in the absence of

Fig. 22. The egg-rolling movement of the greylag goose (after Lorenz and Tinbergen 1938).

external stimuli. The balancing movements, on the other hand, are orienting responses whose presence and course depends on the current stimulus situation—in this case on the position of the egg at a given moment. Hence, the behavior is peripherally controlled, in contrast to the fixed action pattern, which is centrally controlled.

In the example of the egg-rolling behavior of the greylag goose we see a simultaneous intercalation of a fixed action pattern with taxis components that is found in many behavior patterns. Sometimes, both components occur one after the other. Thus, a frog, upon noticing a fly, will first turn toward it and line it up with its body axis; only after this taxis movement comes the fixed action pattern, the flicking out of the tongue, which catches the prey.

The combination of FIXED ACTION PATTERN–TAXIS COMPONENT is not identical with CONSUMMATORY ACT–APPETITIVE BEHAVIOR, as might appear at first glance. The latter two concepts are distinguished with respect to the discharge of action-specific energy and not by their dependence on external stimuli. A certain similarity appears only in that taxis components and appetitive behavior show great variability, while fixed action patterns and consummatory actions are characterized by their stereotype. In comparing the two sets of concepts we can say that consummatory actions are apparently always fixed action patterns (but not all fixed action patterns are consummatory actions), while appetitive behavior can contain taxis components as well as fixed action patterns.

4

Temporal and Hierarchical Organization of Behavior

Behavior patterns occur in a certain temporal order. They may always or primarily occur together, e.g., one after the other in succession, or they may be mutually exclusive and inhibit one another.

4.1. Categorization of Behavior Patterns

Many behavior patterns constitute functional groups that fulfill their biological purpose only when they occur together. These include reaction or action chains, which can be especially observed in social behavior, e.g., during courtship or in ritualized behavior (see Section 8.1.7). Here we find behavior sequences in which the components usually occur in the same order, but where the degree of variability or stereotypy varies from one case to the next. An example of a behavior sequence that was investigated early in the history of ethology, and that has since been frequently cited, is the courtship sequence in the three-spined stickleback. Here, each act of one partner will release the next fixed action pattern in the sequence in the other partner. If the other fish fails to respond, either the previous action is repeated or the behavior sequence is broken off. Similar behavior sequences are known from the courtship behavior pattern of many other animal species. Their biological function will be discussed elsewhere (see Section 8.2.4).

Behavior sequences can consist of units that always occur together. However, there are many movement coordinations that show various levels of integration. They begin at the level of individual muscles: the rhythmic movement of a single spine or ray in the fin of a fish can be caused by the alternating contractions of two antagonistic muscles[1] that are temporally coordinated. To achieve movement of the entire fin the individual movements of the separate rays must be integrated at a higher level. Within a larger framework that comprises locomotion, courtship, nest building, and aggressive behavior, several fins must act together in a meaningful fashion.

[1] ANTAGONISM here refers to the opposing actions of muscles (e.g., flexors and stressors of limbs), hormones (frequently male and female hormones), or nerve cells (e.g., facilitating or inhibiting effects on the function of internal organs). In behavior too there are antagonistic components, e.g., attack and escape.

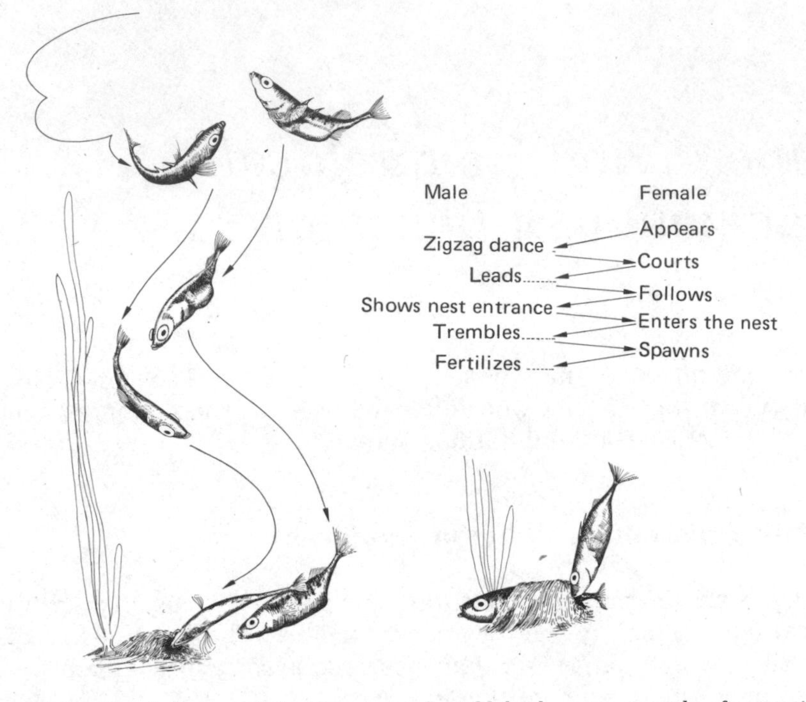

Fig. 23. The mating behavior of the three-spined stickleback as an example of a reaction chain (after Tinbergen 1951).

Fig. 24 (right top). Schematic representation of the relations between male and female three-spined sticklebacks (after Tinbergen 1951).

At a higher level of integration there can also be a temporal order of behavior sequences, although these are rarely rigid. Many times certain behavior elements show a parallel increase and decrease in the frequency of their occurrence within one or a partial functional system (see Section 1.2), or one can detect a certain frequency relationship between the elements. Thus, within the category of reproduction there occur behavior patterns of brood care (nest building, storing of food), courtship, copulation, and care of young in a more or less fixed order. Within these categories individual behavior components, e.g., in the nest-building behaviors of birds, collecting, carrying in of nesting material, and actual construction of the nest can again show a certain temporal order. Behavior sequences that are arranged in some order can be expected to be controlled by higher centers in the central nervous system. This indicates that within the total behavior repertoire of an animal there must exist some kind of "hierarchical" order. This was already seen with respect to motivation (see Section 2.8). The significance of a hierarchical organization of behavior had been recognized early and played a role in the construction of models for the organization of behavior (see Section 4.3).

4.2. Conflict Behavior

In addition to the temporal ordering of behavior elements, the exact opposite can be found: behavior patterns can be mutually exclusive, i.e., the performance of one action can prevent or break off another. Such mutual inhibition is biologically meaningful since it prevents chaos among movement patterns and guarantees the performance of various behavior patterns in their proper sequence. When an animal simultaneously encounters stimuli from potential food and sexual partners, it is ready to respond to both on the basis of its motivational state. Hence, it would be an advantage if the reactions did not occur together but in succession, and one behavior predominated and prevented the occurrence of the other. Such mutual inhibitions are known in many functional systems. It is especially frequent with respect to behavior from *different* functional systems, and is strongly exhibited in agonistic behavior (see Section 8.1.1) between elements of attack and escape.

Normally, the external situation and the motivational state decide clearly which behavior pattern will occur at a given time. Sometimes, however, it is possible that two incompatible behavior tendencies are simultaneously activated and with approximately equal strength. This can lead to conflict situations that are primarily expressed in three ways: ambivalent behavior, redirected behavior, or displacement behavior.

4.2.1. Ambivalent Behavior

Frequently, two opposing or incompatible behavior patterns occur one after another, changing back and forth in quick succession. This occurs when the same object presents stimuli for both behavior categories, e.g., when a rival presents stimuli for attack and escape, when a sexual partner simultaneously presents stimuli for courtship and attack, or when two stimulus sources that elicit opposing reactions are close together. A frequently cited example illustrating this is that of a hungry bird, e.g., a yellowhammer, being next to a higher-ranking bird at a food source. Food and rival present stimuli: the food elicits approach, the rival avoidance. In this situation one can frequently observe incomplete eating movements (rubbing together of mandibles) and intention movements of escape (stretching of the neck and tail movements).

During the course of evolution of the two functional systems both elements can be temporally combined through the process of ritualization (see Section 8.1.3). Behavior patterns with these characteristics are called AMBIVALENT BEHAVIOR. In this sense, courtship behavior patterns can also have ambivalent characteristics, as when they consist of sexual components and those of attack and escape (see Section 8.2.2.2).

4.2.2. Redirected Behavior

A conflict situation can express itself in another form when a behavior pattern is negatively influenced by an opposing tendency and therefore is not directed toward the "correct," i.e., natural, object. This situation is also seen in a conflict between the tendencies to attack and to flee. A well-known example is that of the fighting behavior in herring gulls: if one of the rival combatants is obviously the loser, instead of directing his attack toward his opponent, the bird directs it at a neutral object like a tuft of grass. In a similar manner the biting movements of fighting cichlid fish, which are normally directed at an opponent, can become redirected against the substratum (sand and stones). The behavior pattern is discharged but is reoriented at a substitute object.

Redirected behavior also occurs in social behavior: if an animal is threatened or attacked by a higher-ranking animal (see Section 8.4.6.1), the attack is not directed at the attacker itself but is redirected to a lower-ranking species member instead.

4.2.3. Displacement Behavior

Finally, there are conflict situations in which the "expected" behavior patterns, i.e., those appropriate to the situation, do not occur at all. Instead, a behavior seemingly "nonsensical" in the particular context is exhibited. The most frequently cited example is that of two fighting cocks that suddenly cease their attacks on one another and begin to peck on the ground as if to feed.

Such behaviors are called DISPLACEMENT ACTIVITIES or DISPLACEMENT BEHAVIOR. They are "unexpected" in the sense that they do not fulfill the biologically appropriate function for which they evolved in the context in which they occur.

Displacement behavior always occurs when two incompatible behavior tendencies are activated simultaneously and equally strongly, and hence inhibit one another. Originally it was thought, hence the name, that in a situation of mutual inhibition the "energy" of the inhibited behavior patterns was *displaced* into another functional system and triggered behavior patterns there. If this were so, then the original behavior should occur with lower intensity later on due to the discharge of action-specific energy through the channel of the displacement activity. However, no evidence to support this hypothesis exists to date. Instead, the fighting behavior of cocks, for example, continues with full force after the displacement pecking. These facts and others argue against the original displacement behavior hypothesis.

Fig. 25. Displacement food pecking in fighting domestic cocks (after Tinbergen 1951).

In spite of the large number of investigations and attempts at interpretation that focused especially on displacement behavior, there is still no final agreement as to its nature. The DISINHIBITION HYPOTHESIS is widely recognized as coming closest to an acceptable explanation. It states that the mutual inhibition of two functional systems that are activated equally strongly will allow for the expression of a *third* behavior, although it may be weaker and may have been inhibited until then.

Behavior patterns that are to a certain degree continually ready to be expressed include those involved in feeding, grooming, and preening, and during the breeding period those of nest building and care of young. Displacement activities consist especially of these kinds of behavior. In warm-blooded animals, where the care of feathers or fur is especially important for the maintenance of the insulating layer of air, preening, grooming, snatching, and shaking patterns are frequently "inserted" between other activities. Hence, they are continually ready to be performed and are frequently seen as displacement activities.

The disinhibition hypothesis is supported by the fact that the performance and frequency of the displacement activities are not invariable but seem to fluctuate according to their own specific motivational states. Thus, hungry cocks show more frequent displacement pecking than satiated ones during agonistic encounters, and they may actually eat some seeds. Dirt on the bill leads to increased bill wiping during displacement, e.g., during courtship, and artifical rain leads to increased preening in conflict situations in chaffinches. The disinhibition hypothesis was quantitatively supported in experiments with honeybees that collected nectar. While visiting a food source the bees perform cleaning behavior, especially prior to leaving the feeding place. If one measures the tendency to suck up nectar, which decreases as the stomach is being filled, and the tendency to fly off, which can be recognized by the position of the antennae, then one finds that the cleaning behavior patterns always occur when the opposing tendencies to stay and to fly off are in balance. Cleaning behavior apparently occurs through disinhibition of the constant readiness to perform it due to the mutual inhibition of the tendencies to suck nectar and to fly off. These examples show that displacement activities are not charged by energy from other behaviors, but that their frequency and intensity is dependent on their own motivation.

Displacement behavior not only occurs in conflict situations that are characterized by the mutual inhibition of two other behavior tendencies. They are also known from a number of other conflict situations. This includes situations in which a "goal" is reached too quickly, e.g., when an opponent in a flight flees "unexpectedly," or when an "expected" response does not occur, e.g., when a behavioral sequence breaks off during courtship. An example of the first situation has been described in fighting cormorants, which show sexual displacement activities when the opponent suddenly flees. The second situation exists when two courting sticklebacks perform nest-building activities as displacement when the female does not follow the male in the normal manner to the nest.

In contrast to the "classical" situation for displacement behavior, only one behavior tendency is activated in each case. In spite of this, there seems to be true inhibition when the "appropriate" stimulus is not present or disappears. Then another activated behavior pattern is able to run off because of this disinhibition.

Besides the disinhibition hypothesis there are a number of other attempts to explain displacement behavior. It is possible that various displacement behaviors also have distinct underlying mechanisms. Displacement activities can secondarily acquire signal functions as a result of ritualization and can become a part of social communication (see Section 10.6). Ritualized behavior especially contains courtship and threat behavior patterns.

4.3. Models of Instinctive Behavioral Organization

On several occasions attempts have been made to organize the basic phenomena and regularities of behavior discussed in the previous chapters into some coherent form through some appropriate models. This would give us an overall picture of the phenomena and perhaps explain them to some degree. The organizing principle with respect to behavior involved the concept of INSTINCT. The models that were based on it were called INSTINCT MODELS. There is no more controversial concept in the ethological literature, which is used and understood in so many different ways by various scientists, than the term *instinct*. Since the word also exists in everyday language, where it is used in still another way, and since it often has a certain negative connotation, ethologists use the term less and less. The same problem of definition of *instinct* also exists in the case of *urge* and *drive* (see Section 2.6).[2] On the other hand, composite words like *instinctive behavior pattern*, *instinctive movement* (see the last paragraph in this chapter), and *instinct model* are still widely used, perhaps because they have been a part of ethology from its early beginnings.

The most complete and frequently cited definition of the instinct concept is Tinbergen's. It is given here because it combines many of the terms cited so far, and because it can help in an understanding of the various instinct models. For Tinbergen the term *instinct* means a "hierarchically organized nervous mechanism which is susceptible to certain priming, releasing and directing impulses of internal as well as external origin, and which responds to these impulses by coordinated movements that contribute to the maintenance of the individual and the species."

The most comprehensive instinct model is Tinbergen's. It is based on the "hierarchical" organization of behavior to which we have already referred.

[2] In this connection it is of interest that the term *instinctive* is not as controversial, although it is much debated and equally hard to define. Furthermore, it is today used even more frequently than the term *instinct*. In general, behavior patterns were called *instinctive* when they had an inherited basis. In this sense the word is synonymous with the word *inborn* (see Section 6.1.1).

The model distinguishes between instincts of the first, second, and still lower orders, each of which comprises a subset of several partial behavior patterns or behavior complexes that are components of the next higher level of organization. The highest center is influenced only by internal factors, by hormones in the example given here. It is spontaneously activated, while the lower centers are influenced by internal as well as external factors. The EXCITATORY POTENTIAL from a higher center flows to the next lower one and is normally inhibited by a BLOCK, which is a releasing mechanism (see Section 3.2). This block can be removed by appropriate key stimuli or releasers. The stimulus situation prevailing at a given time determines which behavior

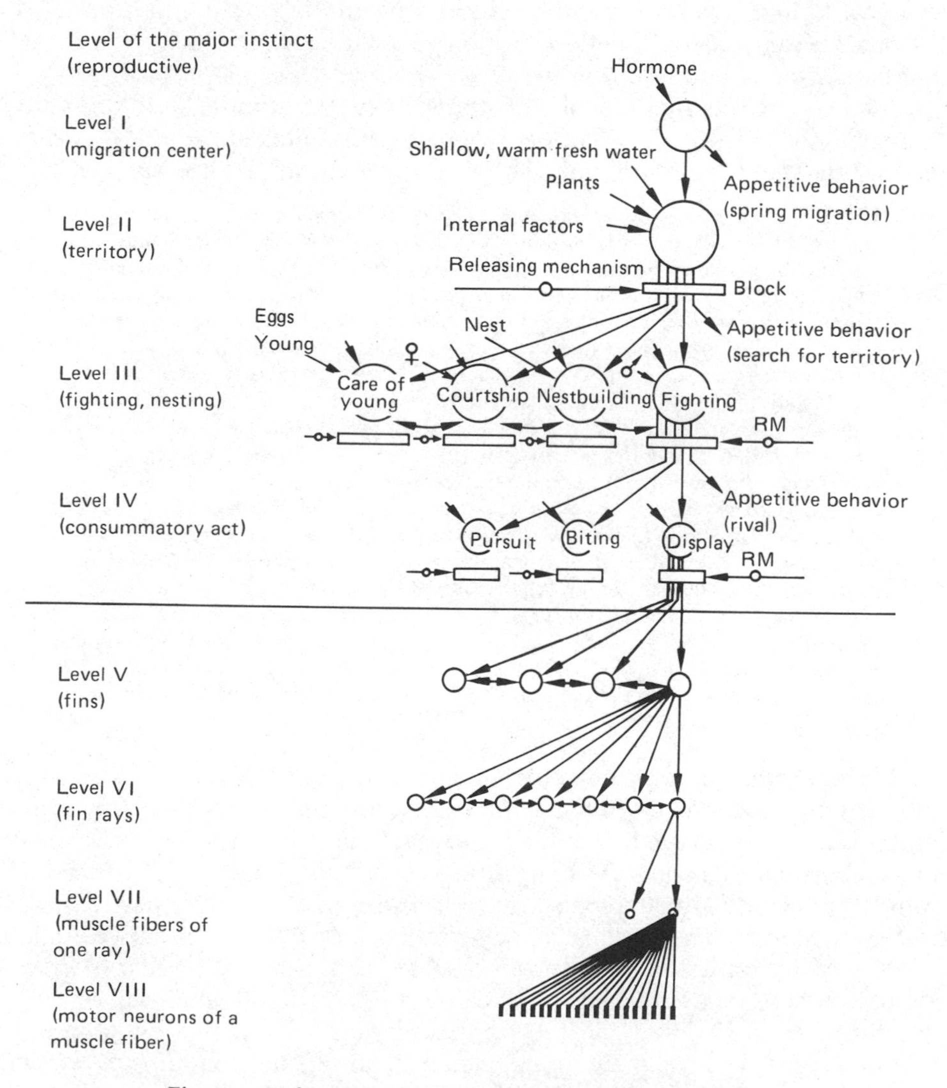

Fig. 26. Tinbergen's instinct model (explanation in text).

patterns of the lower center will be released. If these blocks remain in the absence of an adequate stimulus, and hence the behavior patterns are prevented from being performed, then this excitatory potential flows to those centers that control the appetitive behavior for the particular level or organization. This behavior then is activated until appropriate external stimuli remove the block at this level, which in turn excites the next lower centers, where the same process is repeated. Because of mutual inhibition (see Section 4.2.1) only one particular behavior pattern can occur at a specific time. Whenever several centers are activated simultaneously, then that which is most influenced by other stimuli will pass on its excitatory potential. The lowest level of centers are the consummatory acts in the various functional systems. Below this level there is, instead of mutual inhibition between the various subunits, a coordination such as was shown in the example of the fin of a fish (see Section 4.1).

Tinbergen's instinct model, illustrated here as it applies to the reproductive behavior of the male stickleback, will be explained, since this will enable us to understand the interrelationships among the various centers:

> The hormonal influence, presumably exerted by testosterone, is acting upon the highest center. This center is most probably also influenced by a rise in temperature. These two influences cause the fish to migrate from the seas (or from deep fresh water) into more shallow fresh water. This highest center, which might be called the migration center, seems to have no block. A certain degree of motivation results in migratory behavior, without release by any special set of sign stimuli, which is true appetitive behavior. This appetitive behavior is carried on—the fish migrates—until the sign stimuli, provided by a suitable territory (shallow, warm water and suitable vegetation), act upon the IRM, blocking the reproductive center *sensu stricto*, which might be called "territorial center." The impulses then flow through this center. Here again, the paths to the subordinated centers (fighting, nest building, etc.) are blocked as long as the sign stimuli adequate to these lower levels are not forthcoming. The only open path is that to the appetitive behavior, which consists of swimming around, waiting for either another male to be fought or a female to be courted, or next material to be used in building. If, for instance, fighting is released by the trespassing of a male into the territory, the male swims towards the opponent (appetitive behavior). The opponent must give new, more specific sign stimuli, which will remove a block belonging to one of the consummatory acts (biting, chasing, threatening, etc.) in order to direct the impulse flow to the center of one of these consummatory acts.

A somewhat modified instinct model was conceived by G.P. Baerends. This model also depicts the hierarchical organization of behavior and the mutual inhibition of centers of the same level, but in addition it shows clearly that subordinate centers are often influenced by *several* other centers. This conceptualization illustrates the fact that individual behavior patterns, especially the many elementary behavior patterns, can appear in several functional systems. This applies especially to the behavior or locomotion that is a part of escape, attack, courtship, and prey-catching behavior or that occurs during migration (e.g., between the breeding range and the wintering grounds in migratory birds). However, digging, pecking, biting, and snapping movements can also occur in different functional systems. These important interrelation-

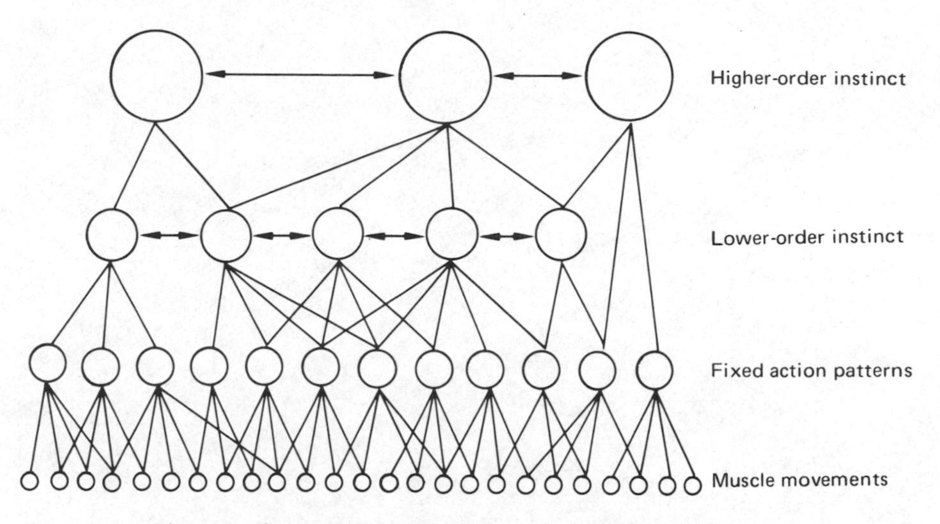

Fig. 27. The instinct model of Baerends (explanation in text).

ships are not accommodated in Tinbergen's model, which depicts only one active subinstinct at a time.

In this connection it must be said that all instinct models were only thought to be "conceptual models," which showed a certain explanatory order for the numerous phenomena and regularities observed in behavior. They can, of course, not provide a true explanation of the underlying processes. In connection with the concept of instinct that is discussed during the description of the instinct models, we must also explain the terms INSTINCTIVE BEHAVIOR PATTERN and INSTINCTIVE MOVEMENT. They are not identical; an *instinctive movement* was equated with the term *fixed action pattern,* referring to the smallest behavior units, while *instinctive behavior pattern* or *sequence* referred to a more complex ordered sequence of fixed action patterns. G. W. Barlow has recently proposed the neutral term MODAL ACTION PATTERN for *instinctive movement* or *fixed action pattern* (modal—referring to mode, manner, or form), since this emphasizes that these behavior patterns can be recognized by their normal appearance, and because it omits any reference to the inherited basis of the behavior that is often misinterpreted as meaning complete independence from environmental influences (see Section 6.1.1).

5

Behavioral Physiology

Many of the facts presented in the previous chapters, especially those referring to the temporal order of behavior and its hierarchical organization, presuppose that somewhere in the organism there should be an integrating mechanism that is capable of ordering and coordinating the various inidvidual behavior elements in some meaningful manner. Principally, there are two steering mechanisms available: the nervous system and hormones. In their coordinating effect within the organism, and between organism and the environment, both show a certain similarity. However, specifically each is characterized by important differences. Hormones act comparatively slowly but can have a longer-lasting effect, while the nervous system acts quickly and in general is responsible for quick and short-term regulating processes. The investigation of both integration systems at the same time offers the opportunity to analyze some of the physiological processes underlying behavior in more detail.

5.1. The Nervous System and Behavior

The basis of behavior in the central nervous system, i.e., the relationship between the behavior of an animal and the function of its nervous system, was of great interest quite early in the history of biology and psychology since such investigations promised to provide conclusions about the control of human behavior as well. At first people were interested in the location of elements in the central nervous system that were responsible for a given behavior—especially in man. There were two opposing viewpoints in the philosophers' discussion: the LOCALIZATION OR CENTER THEORY, and the theory of plasticity. The localization theory was originally based on findings involving the loss of function after a certain part of the brain was destroyed, i.e., each behavior characteristic can be localized in a specific site in the brain. The THEORY OF PLASTICITY, on the other hand, did not accept a strict correspondence of specific functions with distinct areas of the brain. It pointed out the similarity of the constituent elements and the widespread lack of distinct morphological boundaries in the central nervous system. This difference of opinion about the two theories undoubtedly contributed to an early concern with localization of brain function in neuroethological research, and was to be reflected in the choice of research methods.

5.1.1. Methods

Neuroethological research employs the following four methods: ablation, brain stimulation, recording of action potentials, and hormone implantation.

5.1.1.1. Ablation

A first, albeit crude, indication of the localization of control functions of the brain results from the elimination of individual parts of the brain. This can be accomplished by the operative removal of the elements in question (such as by amputation or extirpation). It can involve the cutting of nerve fiber tracts within the brain, or the destruction (lesion) of brain tissue with high-frequency current or laser beams. Recently, local freezing of brain tissue has also been used.

The validity of statements based on ablation studies is, however, limited since this treatment usually involves larger parts of the brain. Furthermore, it is known from other investigations (see Section 5.1.2.1) that even anatomically close brain structures can have quite different functions. In addition, any intervention in the brain has such important consequences for the entire organism that any effects of the ablation are not necessarily the result of the elimination of brain tissue as such, but may be due to more general adverse effects.

For this reason, three phases must be distinguished in experiments involving this method: a postoperative phase, a phase of acute influence on behavior, and, finally, a phase of chronic changes. During the first phase no accurate statements about results of the ablation can be made because of the overall effects of the operation. In the second phase, when the postoperative disruptions have dissipated, the loss or change of behavior elements may permit conclusions that the destroyed brain regions were normally involved in the control of these elements. After some period of time there is frequently a partial or even complete restoration of the originally lost or disrupted behavior. This is an indication that in this third phase some compensation occurs for loss of functions controlled by the ablated brain regions. It seems that another area of the brain has taken over the function of the ablated area.

The first ablation of brain tissue was already done 150 years ago, which makes ablation studies the oldest method of neuroethological research. Because of the disadvantages already referred to, this method has fallen into disuse from time to time. Only recently has there been an increase in the importance of this method because greatly improved techniques have eliminated many of the former sources of error, and the experiments have yielded good results.

5.1.1.2. Brain Stimulation

A very elegant experimental method is that of electric brain stimulation. It is also relatively old and can trace its beginnings back about 100 years. It

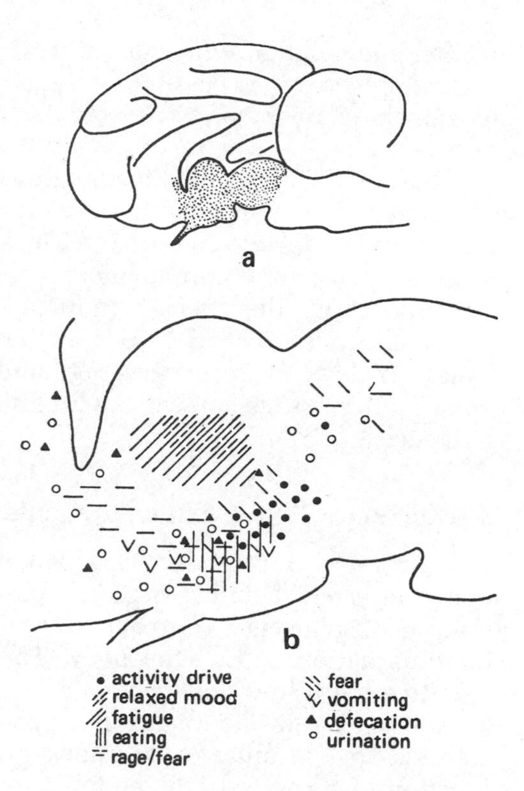

Fig. 28. The brain stimulation experiments by W. R. Hess in the domestic cat: (a) location of the midbrain, (b) position of points within the midbrain in which specific "moods" could be elicited by electrical stimulation (modified after Hess 1957).

reached its peak with the work of W. R. Hess in the 1930s on the brainstem in cats, and with E. von Holst's work in the 1950s on the brainstem in domestic chickens. Today, the results of brain stimulation experiments are available from a large number of vertebrates, primarily birds and mammals, and from several invertebrates, especially insects, crabs, and crayfish.

In stimulating the brain, thin electrodes made of steel or tungsten are used. The wire is insulated with lacquer except for the tip. Thus, the electric current can enter the brain tissue only in a small, limited area. The electrodes are fastened to the skull by a plastic electrode holder inserted in a small hole, and they can be lowered into the brain by turning a screw. The holder can remain permanently in place in the animal's head, while the electrodes, of which several can be inserted through one holder into the brain, frequently are inserted only for a specific experiment. A large surface electrode, a neutral lead, is placed below the skin on top of the skull. By a stepwise lowering of the electrodes, various levels of the brain can be probed in a fairly large area from any given location of the holder. The precise location of the electrode's tip can be ascertained by X-ray pictures or by electrocoagulation. The latter technique makes use of a strong current passed through the electrode, which leaves burnt tissue that identifies the site.

The wires that connect the stimulus apparatus with the electrode holder are kept loose so that the animal can have some freedom of movement. Today, wireless stimulation with transmitters, permitting the animal complete freedom of movement, is possible. Only in some instances, e.g., when the animal is capable of removing the electrodes (as with primates), is it necessary to restrain it.

During stimulation of the brain an alternating current passes from the tip of the electrode in the brain to the neutral lead on top of the brain, or between the tips of two inserted electrodes. The strength of the required current is very low and is measured in milliamperes. To vary the intensity of a stimulus, one can either vary the strength (milliamperes) or the voltage, which usually flucuates between 0 and 2 volts in these experiments. Usually the stimulus voltage is changed since the strength of the current depends on the resistance. This, in turn, is dependent on conditions in the brain that are difficult to influence.

5.1.1.3. Recording of Action Potentials

A reversal of procedures is embodied in the third method of neuroethological research. It is based on the observation that active neurons produce electrical signals as a consequence of electrochemical processes, which pass on information within the body. These potentials can be recorded with sensitive electrodes and they can be measured after suitable amplification. This permits the identification of neurons that are active during specific activities of an animal, and allows statements about the involvement and function of neurons in the brain. It is not always possible to record from a single neuron, but it is possible to separate out the responses of single neurons by appropriate filtering. Electroencephalography is based on a somewhat different principle. Here, electrodes are not lowered into the brain to record an electroencephalogram (EEG); instead, electric potentials produced by neurons in the frontal cortex are recorded by means of surface electrodes attached to the skull. In contrast to single-cell recording, this allows only the recording of activity in larger areas of the brain.

5.1.1.4. Hormone Implantation

Instead of stimulation by electrical impulses in brain stimulation experiments, the brain can also be "stimulated" by the implantation of hormones in crystalline form at specific sites. This method will be discussed in the section on the effects of hormones (see Section 5.2.3.3).

5.1.1.5. Methodological Assumptions

An important consideration for each neuroethological investigation is the selection of an appropriate experimental animal. Valid conclusions about the controlling influence of brain functions can be made only if the behavior of the animal is known in great detail in normal situations. Only this kind of knowledge permits adequate interpretation of the observed behavior. Great importance is attached to the question of whether this behavior is "natural," i.e., if its performance corresponds to its expression under normal circumstances. A good criterion of whether this is the case is whether or not an artificially released behavior is "understood" correctly by a conspecific of the experimental animal, i.e., if it reacts in the usual manner.

Furthermore, the anatomy of the experimental animal's brain should be well known. With respect to brain structure, neuroethologists face certain problems in that the most highly evolved vertebrates, which present us with the most interesting questions because of their many behavior patterns and their plasticity, also possess the most complex brain structures, which are very difficult to analyze experimentally. Hence, in many studies insects and other invertebrates have proved to be more suitable experimental animals. Their CNS contains substantially fewer neurons. The brain of a human contains about 15 billion neurons; that of an insect, on the other hand, "only" from 10,000 to 1 million, which offers many methodological advantages. This is especially true for the recording of action potentials.

5.1.2. Results

5.1.2.1. Localization of Brain Function

Independently of the method of investigation, all neuroethological studies have shown that certain behaviors of an animal are correlated with activity in specific parts of its central nervous system, i.e., that certain parts of the brain are involved at least in part in the performance of specific behavior patterns. When these regions are destroyed, the associated behavior patterns drop out unless they are also under the control of other areas. This correspondence between loss of function because of the destruction of specific parts of the brain is not permanent, since other brain regions can gradually assume control over the behavior in question (see Section 5.1.1.1).

Statements about the localization of brain function are based primarily on ablation studies, according to which the relationship between areas of the brain and behaviors is such a widespread phenomenon that the listing of individual results is superfluous. Many animal species have been so thoroughly investigated that so-call stereotactic atlases exist showing pictures of sections of the brain in three-dimensional coordinates that represent the areas of the brain, which are usually named according to their function.

Brain atlases and information about the spatial distribution of function in the brain provide an important prerequisite for the more precise method of brain stimulation. It is especially suited for a more focused testing of specific hypotheses, because of the placement of electrodes into specified areas of the brain.

Brain stimulation experiments have complemented and advanced the results obtained from ablation studies. They have shown that certain behavior patterns can be artificially released in certain parts of the CNS through electrical stimulation. This is true for motor patterns as well as for vocalizations in birds and insects. In many species it has been possible to release most behavior patterns, and in the domestic chicken almost all behaviors were released through brain stimulation.

The degree of complexity of behavior patterns that can be elicited in this manner varies greatly. They range from very simple movements (e.g., piloerection, pulling back the ears, movements of a limb) to relatively complex sequences of behavior (see Section 4.1), which correspond to the behavior sequences described earlier and which allow the recognition of an orderly sequence of single elements (e.g., escape, fighting, or prey-catching behavior).

With localization of brain centers that control behavior it was found, mainly through electric brain stimulation, that the control functions can be distributed in the form of a mosaic, i.e., that the individual "fields" are spatially limited, and that various distinct behavior patterns can be elicited by two electrodes whose tips are only 0.1 mm apart. Finally, it was found that the same behavior can be elicited on occasion at two or more sites in the brain. This raises the questions of whether the actual centers[1] for the behavior were stimulated, in which the particular movement sequence is "assembled," as it were, or whether higher-order centers that merely activate were involved. It is also possible that some controlling structures that pass on information (impulses) were stimulated. There are many indications that both occurrences are possible. It was found in several bird species (e.g., Cape red-shouldered glossy starling, Steller's jay) that their vocalizations and the associated behavior patterns are organized in a descending system of correlated brain structures that extend from the forebrain to the midbrain.

In these the corresponding reactions can be elicited in any number of places, albeit in different ways, by electrical stimulation. In crickets, various song components (mate attraction, aggressive songs) can be elicited by electric brain stimulation in the thoracic ganglia as well as in the brain. Experiments have shown that the thoracic ganglia possess some independence with respect to the control of the song pattern, while the brain merely sets off the song by an appropriate command, i.e., it "decides" only whether a song is to occur and which type it will be. In crickets the descending system can be mapped one step further with a recording electrode. This investigation has shown that the action potentials in the muscles that provide the signal correspond in a 1 : 1 relationship to the central impulse patterns.

Multiple representation in the brain can be expected in all those cases in which the same movement pattern occurs in more than one functional system, and where it may be associated with several motivations (see Section 2.8).

[1] The concept of *center* in this connection is by no means clearly defined, nor is it possible to define it clearly based on our current state of knowledge. Generally, a CENTER refers to a relatively small group of neurons with the same function. Brain stimulation experiments especially indicate that nerve cells that are functionally correlated, or that are connected in some way, need not be spatially close together. This runs counter to our usual understanding of what a center should be. In spite of these limitations, the term *center* has been usefully applied to these functional units.

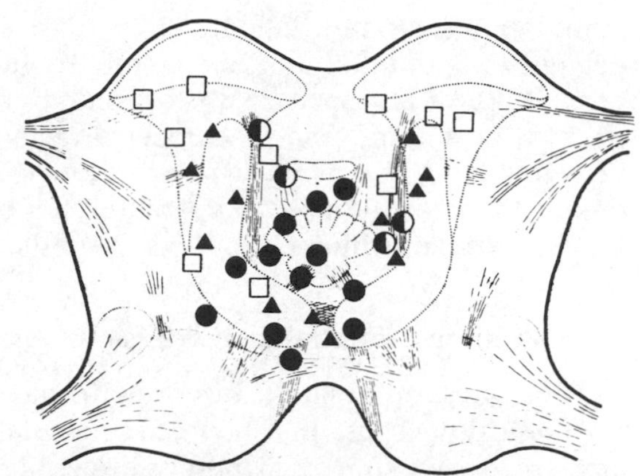

Fig. 29. Section of the frontal plane of the brain of a cricket indicating the regions in which various types of songs can be elicited by electrical stimulation (after Huber 1965).

Hence, sitting down in a chicken can occur in the context of sleeping, incubating, or brooding, or just plain sitting down. A crouching posture when the animal is standing can be a part of a response toward an aerial predator or it can be part of submissive behavior (see Section 8.1.4) toward a higher-ranking rival. These various functions and associations to different functional systems by one single behavior pattern correspond, as was shown in brain stimulation experiments, to their activation in various stimulus sites in the brain.

5.1.2.2. Behavioral Responses to Electrical Brain Stimulation

Behavior patterns that require external stimuli in addition to stimulation of the CNS can be divided into two groups: those released by electrical stimulation alone, and those influenced by brain stimulation but whose occurrence and course depend on external stimuli in the particular case. To the first group belong, among others, the escape behavior of chickens, or the already mentioned songs of crickets. Both behaviors run off after brain stimulation even when the appropriate stimuli (aerial or ground predator or a conspecitic female) are completely absent. The second group in turn encompasses two possibilities: in some instances brain stimulation does not overtly lead to any recognizable result. That the influence is present is indicated when the animal is offered a stimulus to which it has not responded earlier, but which now releases a reaction. Thus, a rooster will attack a stuffed conspecific that had initially been completely ignored but that was suddenly attacked as a "rival." In this case the brain stimulation does not release a behavior directly, but the mood of the animal is changed so that the particular behavior is then more easily elicited by appropriate releasers. In this case the stimulation is comparable in its effect to motivating key stimuli (see Section 3.8). Instances in which running about or searching are released by brain stimulation are somewhat different, i.e., an observable but seemingly "aimless" behavior is exhibited that was frequently (mis)interpreted as general restlessness. If, however, the animal is offered appropriate stimuli like food, a sexual partner, or a rival, then the proper directed behavior patterns occur. Choice tests involving several objects have shown that the reaction is elicited by only one of them, e.g., a rival. Apparently, the brain stimulation elicits a "search for the releasing situation," in other words, a genuine appetitive behavior.

5.1.2.3. Brain Stimulation and Basic Ethological Concepts

Electrical stimulation in the brain has thrown new light on some basic ethological concepts that had been postulated and defined originally on the basis of observations and purely behavioral experiments. The more general results of brain stimulation experiments are best illustrated by the work of E. von Holst and U. v. St. Paul with chickens. These experiments were done some time ago, but they generated so much basic information that can be linked directly to a description of these concepts that it is possible to give examples of many questions. This in spite of a more refined and complex picture that has emerged following more recent investigations.

a. Measuring a Motivation. The motivational state of an animal can be tested in a behavioral experiment by means of models (see Section 1.3). However, it can be also assessed by electrical brain stimulation. The stimuli can be quantified with a stimulus generator that can vary amperage and voltage. In the experimental animal the strength of the elicited reactions can be measured by the latency (i.e., the time elapsed between the application of the electrical stimulus and the onset of the response), the duration of the response, and especially the length of the so-called carry-over (i.e., the time between the cessation of the stimulus and the disappearance of the response). All three parameters depend on the strength of the electrical stimulus (with more intense stimulation the response is stronger, the latency is shorter, and the response lasts longer) as well as on the motivational state of the animal at a given time. The more an animal is already specifically motivated, the weaker the stimuli that are required to elicit the appropriate responses, and the stronger it will react to stimuli of the same strength.

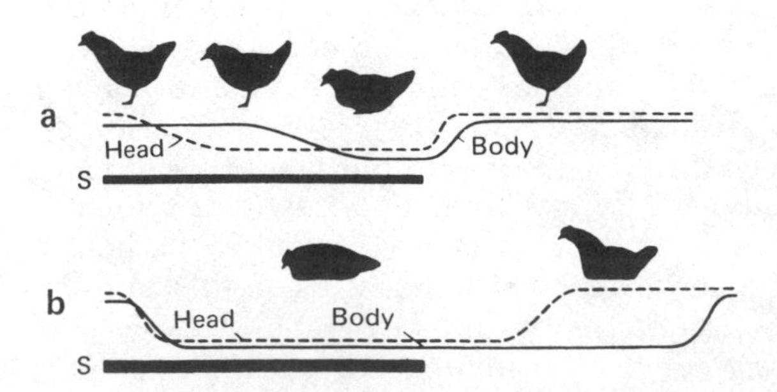

Fig. 30. Effect of two sitting stimuli of different strengths on a quietly standing fowl; pulling in the head and lowering the body are registered separately; the stimulus (S) amounts to 0.5 volt in (a), 0.7 volt in (b) (after von Holst and von St. Paul 1963).

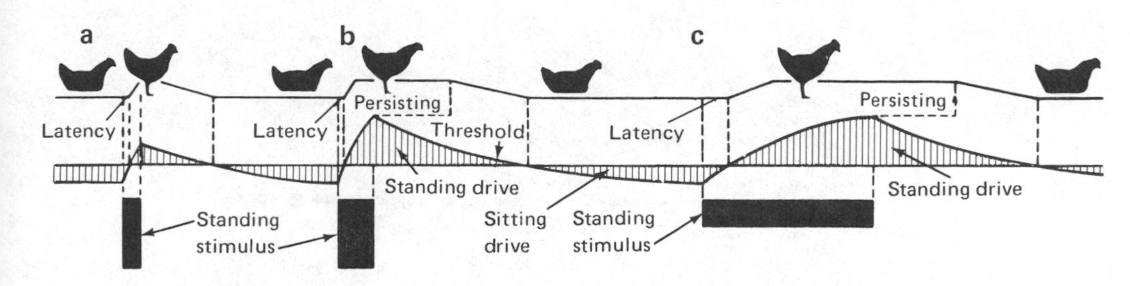

Fig. 31. Sketch to illustrate the interrelation of various parameters of the behavior (explanation in text) (after von Holst and von St. Paul 1963).

These experiments have yielded, among others, the following results. The internal motivation of externally identical behaviors may vary in intensity: a sitting hen, motivated by sleep or incubation, will sit very "tight," while one that following some other emerging motivation is just about ready to stand up "barely" sits. In this case one can vary the voltage and measure the strength of the tendency to sit: the stronger the motivation to sit, the stronger the stimuli that are required to cause the animal to get up. As in experiment investigating the releasing effect on sensory stimuli (see Section 2.2), one speaks here also about the releasing threshold or about the threshold stimulus current for a particular behavior pattern. Stimuli that remain below this value are called SUBTHRESHOLD STIMULI. This statement can also be rephrased to read that the minimal stimulus current that elicits a given reaction (here, standing up) can vary even when the external stituation (here, a sitting hen) is the same. Single elements of a behavior sequence can also have different thresholds for release at the same time: if one stimulates a certain point in the brainstem, then different behavior elements appear with weak stimulation than when stronger stimulation ia applied. Usually bill shaking, tongue movements, and regurgitation movements follow one another during the so-called disgust reaction in a chicken. These behavior patterns can be elicited at the same site in the brain by weak, medium, and strong stimulation. If a strong stimulus is applied first, then the first two behavior elements are skipped and regurgitation appears at once. Similar results were obtained for other behavior sequences.

Behavior patterns can also be "fatigued" by repeated brain stimulation at short intervals, i.e., they become gradually weaker, the latency increases, and the response duration decreases until they cease all together. The behavior can then be elicited only after some suitable interval before the next stimulation or if a much stronger stimulus is applied. This is referred to as a "pumping dry" of a reaction.

Such a decrease in the strength of a reaction differs for various functional systems and movement patterns. This is reminiscent of the action-specific fatigue of behavior patterns (see Section 2.3). The smallest behavior units, i.e., simple movement patterns that belong to several motivational systems (see Section 2.8), can be elicited for a long time without showing signs of fatigue.

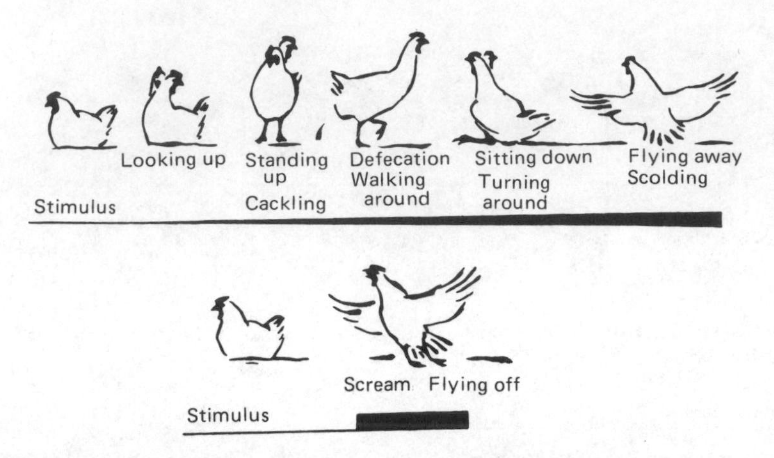

Fig. 32. Behavior sequence of fleeing from a ground enemy with slowly increasing stimulation of the field (top), and the reaction to a sudden, strong stimulus (bottom). In the latter instance the intermediate patterns are omitted and the behavior with the highest threshold occurs at once (after von Holst and von St. Paul 1963).

b. The Interaction or Combination of Behavior Patterns. Especially interesting results were obtained by brain stimulation experiments in which *two* points in the brainstem were stimulated simultaneously. Here, the two stimulus sites can control or activate either the same or two different behaviors.

During stimulation at sites for two different behavior patterns there are again two possibilities: both stimuli can be either of the same strength or of different strengths. In the first instance both reactions, e.g., eating and looking up, rapidly alternate, or neither of these behavior patterns is observed but a completely different one occurs. Alternation is reminiscent of the ambivalent behaviors seen in overt behavior without brain stimulation (see Section 4.2.1), and the appearance of an entirely new reaction seems like a displacement activity.

If one stimulus field is intensely stimulated and the second less so, only

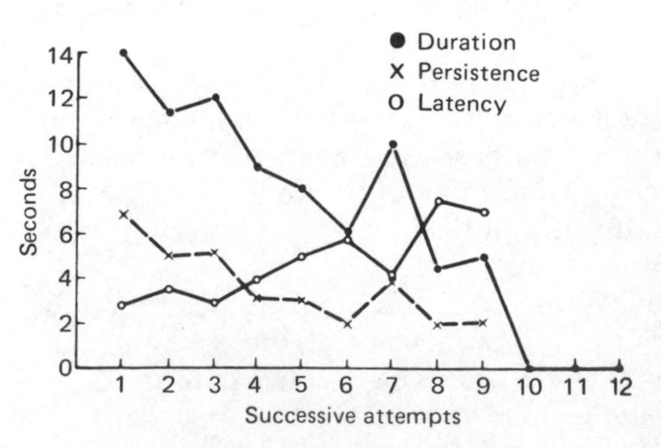

Fig. 33. A stimulus field for clucking was "pumped dry" by a series of 12 stimulations, each of 10 seconds' duration (with intervals of 10 seconds between stimulations) (after von Holst and von St. Paul 1963).

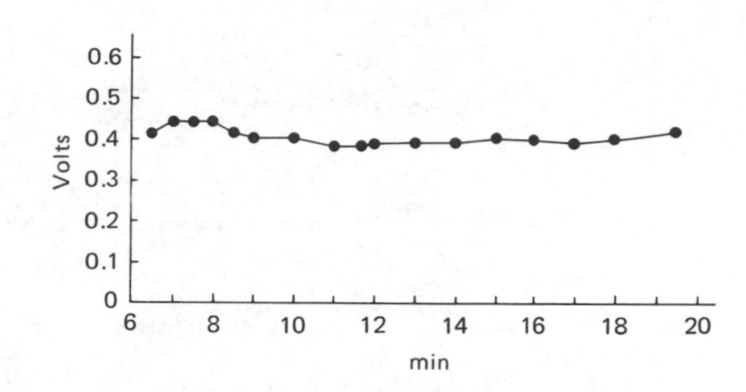

Fig. 34. The so-called low-intensity cackling as an example of a (very simple) behavior element, which in contrast to "regular cackling" can be released continuously in about the same way (after von St. Paul 1964).

the reaction from the first stimulus site appears. The response from the second site is suppressed. This indicates that there must be inhibiting cross-connections between nerve elements underlying the various behavior patterns. An interesting phenomenon occurs in a situation where the difference between stimuli is small: at first, only the response to the stronger stimulus occurs, but suddenly the suppressed behavior pattern can burst forth. For example, if cackling behavior is elicited in a chicken, but it is suppressed by a stronger stimulus for sitting, then cackling suddenly appears in full strength. When two stimulus sites for the *same* behavior pattern are stimulated one can observe behavior that corresponds to the LAW OF HETEROGENEOUS SUMMATION. If both points are stimulated separately with intermediate intensity, then a reaction of intermediate strength is obtained. If, on the other hand, one stimulates both points simulataneously, then the strength of the response is clearly increased (Figure 35b). Hence, the effects of stimulation are additive, i.e., the excitation from both sites must combine somewhere. In the same way separate stimulation of two points can be subthreshold, i.e., no response occurs, but the reaction appears as soon as both stimuli are presented together (Figure 35d). Finally, if one "fatigues" a particular reaction through repeated stimulation in one site, the same reaction can again be elicited at another site that also controls the reaction (Figure 35c).

This is another proof that action-specific fatigue of a behavior (see Section 2.3) is not simply motor fatigue. The release of the behavior must already take place *before* the excitation from the two stimulus sites combines, since otherwise the effect from one site would also be evident from the other site. This is called CENTRAL–LOCAL ADAPTATION (Figure 35e).

5.1.2.4. Central Programming of Behavior Patterns

Even more sensitive than electrical brain stimulation is the recording of action potentials in the CNS, provided that it is possible to record from a single brain cell. In vertebrates such attempts meet with great methodological difficulties (see Section 5.1.2.1). Nevertheless, some results are available. In rats it was possible to demonstrate the existence of single neurons that showed an increase in rate of discharge when the animal engaged in specific activities,

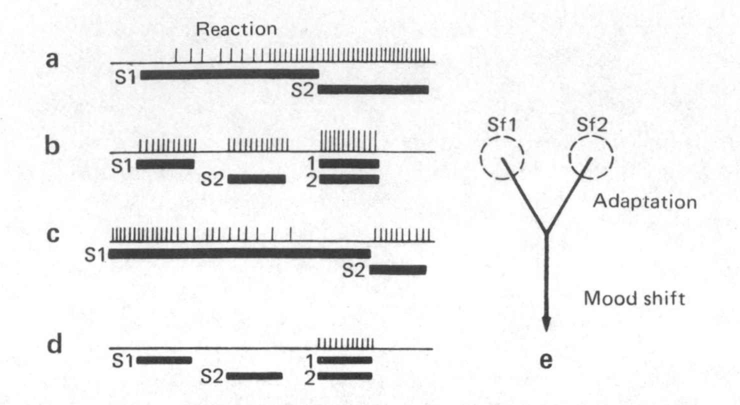

Fig. 35. Scheme to explain various combinations in the stimulation of two fields for the same behavior. Based on a case in which the two stimulation fields lay in the right and left brainstem halves, laterally separated by a distance of about 5 mm, and gave "pure" clucking. To be certain of excluding a possible mutual physical interaction due to voltages produced in the tissue by the two pair of electrodes, when both fields were stimulated simultaneously, rhythmically interrupted stimulation was used in such a way that when both fields were stimulated the stimulation in one fell during the pauses in the other (S = stimulus, Sf = stimulus field; further explanation in text) (after von Holst and von St. Paul 1963).

e.g., eating, sniffing, or movements of the vibrissae. After cessation of the activity the electrical activity decreased again. A similar increase in impulse rate was found in certain cells in the midbrain of the cat as a reaction to attack-releasing stimuli, such as the presence of a male rival. In several fishes, e.g., wrasses and sunfish, specific action potentials can be recorded from single neurons during specific behavioral events such as approach, chewing, and swallowing of prey.

Much more extensive recordings have been made from the central nervous system of invertebrates. Insects and crustaceans are especially suited for such work. They have a comparatively small number of cells in the CNS (see Section 5.1.2.1) and have the advantage that many of their behavior patterns show a great constancy of form (see Section 3.9.1) and can therefore be readily quantified.

In insects not only was it possible to correlate impulses from specific neurons with a specific movement, but entire impulse patterns that matched the appropriate movement patterns could be recorded. Most suitable for these investigations are simple, rhythmic movements, e.g., the raising and lowering of wings during flight or the stridulating movements of the wings or legs in chirping males.

In this manner one can let desert locusts "fly" in place in a wind tunnel by attaching them to a harness on a rod. As soon as the animal loses contact with the ground and senses an air current with its sensory hairs on the forehead and the top of its head, it begins to fly. The wings of one segment beat in synchrony, while the anterior and posterior wings on the same side of the body are temporally slightly out of phase. In contrast to the majority of insects, locusts have "direct" flight muscles, i.e., elevators and depressors that are attached to the base of the wing and contract alternately and thus provide the power for the up-and-down beat.

If one now records from the motor neurons in the thoracic ganglion in such a "flying" locust, which transmit the "instructions" for the muscle contractions, then one obtains not only alternating impulses from elevators and depressors but also synchronous impulses for the two elevators of the two wings of the same segment, and impulses following one another with a slight delay in the elevators of the wings on the same side. This rhythmic pattern is completely "natural," i.e., it corresponds precisely to the normal wing movements, which have been recorded on film and which can be precisely analyzed.

Information about the origin of this neural pacemaker is provided by experiments in which the sensory information emanating in the flight muscles and leading to the CNS is interrupted in successive steps. Thus, it is possible to hold down the wings artificially and to prevent any movement, or one can cut the sensory nerves that lead from the stretch receptors (see Section 2.5) in the wings and thus eliminate any feedback. Even under these circumstances the normal, rhythmic impulse patterns coming from the motor nerves can be detected. Only the frequency of impulses is reduced in this case. Finally, it was possible to record the normal motor pattern for the flying movements from a completely isolated thoracic ganglion.

The basic rhythm in the CNS in this case, as is known from other experiments, originates in the thoracic ganglion. The "generator" that is responsible for this functions with complete autonomy, i.e., the locust can produce its flying pattern even if it receives no feedback about the position of its wings at a given time.

One can also record rhythmic impulse patterns from the flight muscles of a chirping cricket that correspond exactly to the movement patterns with which the various song types (see Section 5.1.2.1) were produced by stridulating wing movements. The impulse pattern is maintained even here if—in this case through amputation of the forelegs, which contain the auditory organs in the tibia—sensory input is excluded.

These results support what will be discussed in another context (see Section 6.1.3), namely, that the CNS contains ready "programs," i.e., complete impulse patterns that produce a particular movement sequence in the appropriate muscles. This production of orderly impulse patterns without appropriate sensory input, i.e., without corresponding stimulus input from the sense organs, is known as CENTRAL CONTROL or CENTRAL PROGRAMMING. It has since been possible to demonstrate this for a large number of behavior patterns in various species.

5.1.2.5. Concluding Remarks

Overall, the various neuroethological methods employed in various experiments have brought out two important general results: they have led to better insight into many basic phenomena and into the regularity of behavior, and they have shown that there are concrete relationships between specific areas of the CNS and specific behavior characteristics.

In spite of this a word of caution is appropriate. The large number of currently available results should not obscure the fact that, especially in the area of neuroethology, the number of unanswered questions is still surprisingly large. This is especially true at the neuronal level, i.e., with respect to the possible function and significance of individual nerve cells with respect to behavior, where our knowledge about the parts of the brain recognized as relevant centers is still scant and has not even touched the basic questions about the effective mechanisms.

The spatial correlation between nervous system and behavior, as important as these investigations were initially for methodological reasons and for the many-faceted theoretical considerations (see Section 5.1), will not be the central and final goal of neuroethological research. Today, it seems more important to investigate the way in which the CNS works in its totality with respect to the control of behavior patterns. Hence, the coordination between and interaction of its various parts and the possible connections to the sensory input (i.e., messages from the sense organs) and the motoric output (i.e., the "orders" to the effector organs) are increasingly the focus of general interest.

5.2. Hormones and Behavior

The second integrating system of the organism, the hormone system, works much more slowly, but its effect is longer lasting. Thus, it is especially suitable for the regulation of behavior systems that are subject to long-term fluctuations. This includes, first of all, the functional system of reproduction. The most obvious influences of hormones on behavior are known from the realm of sexual behavior and care of young. They are also evident to a lesser degree in other areas.

5.2.1. Methods

Ethoendocrinological research employs the several methods, discussed here in order of increasing ability to warrant definite conclusions.

5.2.1.1. Investigations of Hormone-Secreting Glands and Hormone Levels

The first indications of the possible hormonal control of a behavior are given when endocrine glands, which produce a specific hormone, show seasonal fluctuations in size and activity and when at the same time one can observe differences in the behavioral activities of the animal. This is true primarily for sexual behavior. Birds and mammals have been investigated most intensely. With the exception of some tropical species (see Section 8.2.2.4) and domesticated animals living under artifical conditions, the gonads, which are the most important sites for the production of sexual hormones, undergo seasonal changes in size that may be substantial. The gonads regress toward the end of the breeding season and enlarge again toward the beginning of the next one. Thus, the volume of the testes fluctuates in the jackdaw between 4 and 1200 mm^3 during the course of the year, and in the European blackbird from 1 to 400 mm^3. The testes of the ring-necked pheasant attain 1200% of their winter weight in the spring. Fluctuations of sexual activity are temporally coupled with these differences in the size of the gonads. Hence, when the gonads are the smallest, sexual behavior is usually absent, and it will only appear when the gonads again increase in size. In some bird species there is a very close correlation between testes size and the intensity of courtship behavior. Histological and biochemical investigations indicate a direct connection between gonadal activity and behavior.

A more immediate correlation with behavior can be detected by assessment of blood hormone levels with the use of radioimmunological assays in which it is possible to measure the concentration of individual hormones with the aid of radioactively marked steroids and specific hormone antibodies in the range of picograms (10^{-12} g). This method was only recently developed and has been applied to large animals with success. There are still problems with smaller species because the amount of blood that can be safely withdrawn is too small for an assay.

Radioimmunological assays are especially useful for the determination of the "reverse" influence, i.e., for the investigation of effects that certain behavior patterns have on the concentration of certain hormones in the blood plasma (see Section 5.2.4). This method, among others, has been used to detect changes in sexual behavior (courtship and copulation).

5.2.1.2. Operation

The oldest method for assessing the role of hormones in all areas of biology including behavior is the removal of individual endocrine glands. If a behavior disappears, the conclusion seems warranted that the hormone of the gland in question is responsible for its control. Such experiments contain a source of error, as the operative procedure itself can lead to general damage of the organism. Thus the absence of individual behavior patterns cannot be traced without question solely to the absence of a specific hormone. To guard against this possibility, sham operations under identical conditions must be performed.

5.2.1.3. Hormone Substitution

The best evidence for direct influence is obtained if a behavior that was not shown or is no longer performed is reinstated after the artificial administration of hormones. A long-term effect can be achieved with the implantation of hormones in crystalline form that only gradually diffuse and enter the bloodstream.

5.2.2. Influences of Hormones

Most experimental investigations on the effect of hormones were done on the sexual behavior of vertebrates. Numerous results are available for mammals, especially from rats and rhesus monkeys.

The male sex hormones of vertebrates are called ANDROGENS. They are primarily produced in the testes. The most important androgen is testosterone. In females, there are two important groups of hormones, the ESTROGENS and the GESTAGENS. The estrogens are primarily in the ovaries. They control the maturation of follicles and are responsible for the development and maintenance of female sexual characteristics. The most important one is the follicular hormone estradiol. The gestagens are also involved in the regulation of sexual processes, but their effects are evident much later than those of the estrogens, and they influence primarily those segments of the reproductive cycle that follow copulation. In mammals this includes the preparation of the uterus for the implantation of the ovum. The most important gestagen is the hormone progesterone, which is produced by the corpus luteum.

The control of behavior by hormones can be most clearly shown in the male sex, since there is normally only one group of hormones involved in the regulation of behavior.

5.2.2.1. Male Sexual Behavior

a. Castration. The removal of the testes in all vertebrates that have been investigated results in a widespread if not complete loss of all sexual behavior. The following regular events take place: in most instances sexual behavior does not drop out at once, but wanes gradually. The duration of the sexual aftereffects varies with the species and spans a period of up to five months in rats and zebra finches. In dogs, the aftereffects can last up to five years. In general, the castration effects appear faster in younger individuals than in older ones (see Section 5.2.2.7). Furthermore, all behavior patterns of the sexual functional system do not disappear at once. The ability to copulate is lost first. This is followed by behavior patterns with increasingly lower thresholds. Courtship initiations and other elements with relatively weak sexual motivation are retained the longest. Finally, it is known that individual, low-threshold sexual behavior patterns can, in some animals, occasionally be activated at all times. Here the suspicion is that they are directly controlled by the pituitary gland (see Section 5.2.2.4), or that the function of the sex hormones is partly taken over by hormones secreted by the adrenal glands.

Instead of operative castration, it is now possible to castrate chemically. This involves the use of so-called antiandrogens (e.g., cyproterone), which prevent the effects of male sexual hormones by blocking the androgen receptors. The consequences for the performance of sexual behavior are essentially the same as when the testes are removed. Methodologically, chemical castration has the advantage that its effects are reversible; after chemical decomposition of the antiandrogens over a few weeks or months, complete sexual activity is restored.

Another "nonbloody," but irreversible, form of castration that has been used in recent times is the destruction of gonadal tissue by highly concentrated X rays.

b. Substitution. One can positively influence male sexual behavior by the artificial administration of androgens. These experiments can be carried out in various groups of individuals: in intact males outside of the normal reproductive period, in castrated animals, in young animals, and finally, in females. In this way, it is possible to cause birds to sing in midwinter by the application of testosterone, and sticklebacks will begin to court. In castrated animals, one can gradually achieve complete compensation for the eliminated behavior. During this process, individual sexual elements reappear in the reverse order in which they dropped out after the operation.

The effect of testosterone is especially pronounced in young animals. Thus, chicks crow when only a few days old, and male rats exhibit their first copulatory behavior at 14 days of age. Young male dogs show raised-leg urinations in the manner of adult dogs after the injection of testosterone. In females, similar effects can be achieved. Female canaries begin to sing after treatment with testosterone, hens show male courtship behavior patterns and crow, and female rats attempt to mount others.

The appearance of male sexual behavior in females shows that often both sexes possess male and female behavior patterns, but that one is usually absent because the appropriate hormone is lacking or is not effective because of the antagonisitc influence of the opposite sex hormone (see Section 5.2.2.6).

5.2.2.2. Female Sexual Behavior

With regard to the dependence of female sexual behavior on hormones, the results are less uniform. This is due to the participation of two separate hormone groups. Such a direct dependence of just one hormone, as is the case in the male, seems to occur only rarely in the female.

Very clear-cut results are obtained even here with castration. The effect is even more drastic than in males, since sexual behavior in the female does not gradually disappear during the course of weeks and months, but usually drops out at once. Its restoration occurs in various ways. In rabbits and cats, receptivity is restored completely through the administration of estrogens. In other mammals (rat, sheep), both hormone groups are required. Here, the application of estradiol elicits only the first stages of sexual behavior, and a complete development of sexual behavior requires the addition of progesterone. In many cases, the full effect is only assured when estrogens and gestagens are administered in a specific temporal order and in dosages that are adjusted to each other. In contrast to androgens, the effect of female sex hormones is restricted primarily to this sex; in males, it is usually not possible to elicit sexual behavior by the administration of estrogens.

5.2.2.3. The Effects of Sexual Hormones on Other Behavior Systems

Besides sexual behavior in its normal sense, sex hormones influence a number of other behavior systems, most of which, however, are involved with reproduction in one way or another. The influence on aggressive behavior is especially strong (see Section 8.1.9.5). Nest building is also under hormonal control, but the type of hormones that are involved depend on which sex builds the nest. If only the male builds, as is the case with the black-fronted diok and in several fishes (e.g., stickleback, threadfin), then nest building is activated by testosterone. If the female builds, then female sex hormones are involved. However, their involvement is variable: in mice, nest building is controlled by progesterone; in ringdoves and parrots of the genus *Agapornis*, by estrogens; in the Siberian hamster, both sex hormones together control sexual behavior; and in rabbits, a change in the relative amount of both hormones in relation to each other is important.

Experiments in rats and chicks have shown that sex hormones can also influence the ability to learn. In both instances, the learning performance was improved after treatment with testosterone with respect to some criteria. Such a linkage is biologically meaningful in that during the breeding season, additional details in the animal's *Umwelt* (comprising the effective stimuli important to the species) need to be known. Territorial boundaries, the building of the nest and its location, personal identifying marks of territorial neighbors, the mate or the young are important, and thus the need for increased learning ability is made during this time.

Finally, the extent of the total activity of an animal can be positively correlated with the sex hormones. However, generally valid statements on the basis of currently available results are not yet possible. It does seem appropriate that this kind of change should take place in view of the increased demands made on animals during the reproductive period.

5.2.2.4. Other Hormones

Other than sex hormones, there are comparatively few and relatively nonuniform indications of specific hormone effects. The hormones of the adrenal cortex seem to have effects similar to those of sex hormones, and they are like them in chemical structure.

The most important organ in the body responsible for the control of hormones is the pituitary gland at the base of the brain. It secretes two kinds of hormones: systemic hormones, which reach the effectors directly, and so-called glandotropic hormones, which affect the activity of other hormone-secreting glands and which have only an indirect effect. This dual effect is also seen in the area of behavior; some pituitary hormones can directly affect certain functional systems of behavior (e.g., care of young, aggression). The hormone prolactin controls behavior patterns involved in the care of young (e.g., in pigeons, hens, and various species of fish) and, at the same time, facilitates the development of certain food-producing organs [mammary glands in mammals, lining of the crop for the production of "crop milk" in pigeons, and mucous secretions on the skin of discus fish (see Section 8.3.1)]. Other pituitary hormones are effective only by regulating the amount of secretion of other hormones, e.g., the sex hormones. In some instances, both effects occur; hence, the two pituitary hormones, FSH (follicle-stimulating hormone) and LH (luteinizing hormone), control, on the one hand, the secretion of gonadal hormones (gonadotropic effect), and hence behavior; on the other hand, they seem to have in some instances a direct influence on several behavior patterns. For example, in some cichlid fish, even castrated males show completely normal sexual behavior. Several results indicate that pituitary hormones control this behavior, while the effect of androgens is comparatively small.

Some "pituitary hormones" are produced not by the pituitary gland but in the hypothalamus, and are merely stored in the pituitary gland.

5.2.2.5. Hormones in Invertebrates

Of all invertebrates, crabs and insects have been the most thoroughly studied. Here, too, the influence of hormones is most clearly demonstrated in the realm of sexual behavior. However, there is a fundamental difference in contrast to vertebrates in that the gonads do not produce hormones, i.e., they are always produced outside of the gonads. In male crabs, sexual behavior is controlled by hormones that are produced in the so-called androgen glands located along the sperm duct (vas deferens). In cockroaches, a hormone secreted by the corpora cardiaca, a paired gland near the brain, is responsible for the disinhibition of copulatory movements that are controlled by the subesophageal ganglion (ganglion pharyngeum inferius). Female grasshoppers and fruitflies (*Drosophila* sp.) are capable of copulating with a male only under the effect of a hormone secreted in the corpora allata, a similar pair of glands.

In addition to sexual behavior, hormones can also control larval cocoon-spinning behavior and general activity. Overall, experimental results from invertebrates are too few to support general conclusions.

5.2.2.6. Interaction of Several Hormones

All of the examples presented so far show that the control of certain behavior patterns is not exercised by only one, but instead by the interaction of several hormones. An important exception is male sexual behavior.

Generally speaking, hormones can act synergistically or antagonistically, i.e., can inhibit one another. Synergistic relationships exist between gonadal hormones and those produced in the adrenal cortex (see Section 5.2.2.4). Antagonistic hormones usually consist of male and female sex hormones, but not always. Both have one important antagonist in the pituitary hormone prolactin, which, when secreted in larger quantities, controls the transition from the courtship phase to the care of young.

The relationships between pituitary and systemic hormones are especially abundant. On the other hand, there is the regulating function of the pituitary gland on the activity of other hormone-secreting glands (see Section 5.2.2.4), and on the other hand, there is the collaboration between hormones that are directly secreted into the body and the hormones of other glands. In the paradise fish, the construction of its species-typical foam nest depends on the separate actions of two hormones, each of which has one specific effect: testosterone increases the frequency with which the animals swim to the surface for air, and the pituitary hormone prolactin activates the secretion of mucous in the mouth, which makes possible the production of air bubbles that constitute the building material for the nest.

In a few individual instances, there is a cooperation between male and female sex hormones, although their effects are usually antagonistic. Hence, in the courtship behavior of the ringdove, some behavior elements (chasing, bowing) are controlled in the usual manner by testosterone, while the showing of the nest is controlled by the female hormone estradiol, which is probably also produced in the testes.

In general, it can be said that a particular hormone does not have only *one* effect, but that the influence of hormones on behavior is usually dependent on the interaction of several. Hence, even a small change in the relative proportions of the hormones to each other can have an effect.

5.2.2.7. The Effect of Experience

Quite aside from the complex reciprocal interactions between various hormones, the influence of the hormonal system should not be viewed in isolation. There are also similar relationships with other mechanisms that control behavior. This becomes obvious when one compares species. Such comparisons show that behavior patterns in some species are activated by hormones, while in others they are independent of the hormone level.Thus, various behavior patterns of care of young (licking, keeping warm, retrieving) are performed in rats even by young or castrated animals. In some birds and fishes too, various brood-care and nest-building behaviors are largely independent of hormones. They are probably controlled only by the nervous system. The relative contribution of neural and hormonal control apparently varies greatly between species and functional systems.

Beyond that, intraspecific differences in the degree of hormonal dependence are known. They are due primarily to experience. This conclusion follows from the fact that castration usually has a stronger effect in young animals (see Section 5.2.2.1), and that the artificial administration of hormones is weaker than in older animals.

Thus, estrogens elicit nest box visits in female budgerigars more rapidly in experienced females than in animals that have never incubated before. In ringdoves, the activation of care-of-young behavior by progesterone and that of feeding behavior through prolactin appears faster in experienced females than in inexperienced ones. Similar results are available from fishes.

A comparison of this phenomenon within the class of mammals allows us to discover one additional regularity: while the differences in the effects of hormones that depend on experience are relatively small in rats and other rodents, they are greatest in the most highly evolved mammals. In some instances, almost complete compensation for the loss of hormones can be seen; thus, castrated male cats show decreased but contnuous sexual activity if they have had sexual experience prior to castration. In similarly treated animals that have had no experience, sexual behavior disappears as usual. Similar results exist for male dogs, and in several monkey species the castration of adult males has hardly any effect on sexual behavior.

It is possible that there exists a phylogenetic trend in the direction of increasing independence of sexual behavior from sex hormones. Such a tendency would be in agreement with the increasing importance of the cerebral cortex in vertebrates.

5.2.3. Effects of Hormones

The manner in which the influence of hormones is effected is just as varied as are the expression and extent of hormonal control. In general, hormone levels merely influence the motivation (see Section 2.6) of behavior, and hence determine whether or not and how frequently the behavior is exhibited, but not the actual performance of a particular movement. This is exclusively coordinated in the central nervous system.

According to our current state of knowledge, there are at least five different direct and indirect ways in which hormones can be effective. They are not mutually exclusive. Instead, hormones, as a rule, affect behavior in various ways at the same time.

5.2.3.1. Influences on the Sender of Signals

Hormones can improve the effectiveness of releasers and hence they can indirectly elicit stronger reactions. The red coloration of the underside of the stickleback male, which is an important releaser for aggressive behavior, develops under the influence of male sex hormones. Other secondary sex characteristics[2] that have a communicative function in the agonistic and sexual functional systems—e.g., the rooster's comb, the nuptial feathering of birds, or the sexual perineal swelling in female chimpanzees—are all influenced by sex hormones or through the gonadotropic hormones of the pituitary gland. In the area of acoustics, there is a possible parallel in the "breaking" of the voice in several bird species. Finally, there is the production of odors in female insects that attract males and stimulate them sexually, which also depends on hormones, and which is controlled by the *corpora allata* (see Section 8.2.2.1).

[2] SECONDARY SEX CHARACTERISTICS are those bodily features that are present in both sexes but are not actual reproductive organs (e.g., body size, colorction, or shape). Many of these characteristics can be important releasers in the sexual functional system.

Changes of releasers that are controlled by hormones either occur only once in the life of an animal (the differentiation of secondary sexual characteristics upon reaching sexual maturity) or they occur annually (nuptial coloration), or as perineal swellings that occur at intervals of several weeks.

5.2.3.2. Influences on the Receptor

In another way, hormones affect the performance of sensory organs and thus change the perceived and transmitted messages. Thus, testosterone increases the sensitivity of the surface of the penis in male rats and leads to stronger feedback of tactile stimulation, which in turn activates sexual behavior. The sensitivity of the brood patch in many birds is increased by the hormone prolactin, which helps to activate brooding behavior.

5.2.3.3. Activation of Neuron Groups

Besides the interaction of several hormones, there are also reciprocal relationships between hormones and individual parts of the brain. Their investigation can contribute substantially to an understanding of the mode of operation of hormones.

A first method for the investigation of this interaction is the injection of radioactively marked sexual hormones into castrated animals. This results in an obvious concentration of radioactivity in the cell nuclei of the brain, which varies in different parts of the brain. In male ringdoves, radioactivity following the injection of testosterone is strongest in the hypothalamus, and less so in the cerebrum. In female rats treated with radioactive estradiol, the strongest concentration was also found in the hypothalamus, an intermediate concentration was found in the hippocampal region, and the least activity was found in the cortex. Radioactivity reaches a peak about 1–2 hours after injection, and ceases about 2 hours later. It appears as if individual areas of the brain have the capacity to take up hormones that have reached them via the bloodstream and to maintain them in the cells and cell nuclei for some time.

A second method of investigation works quite differently. It is similar in some ways to electric brain stimulation (see Section 5.1.1.2), but instead of electrodes, minute quantities of hormone crystals are placed directly into the brain, where they gradually dissolve into the surrounding regions. By means of such implantations it has been possible to activate individual behavior patterns. These effects are achieved only in rather specific areas of the brain, as was shown by "mapping" large areas of the brain. Thus, in doves, various components of courtship behavior are released in separate stimulus sites by the implantation of testosterone crystals. In castrated roosters, testosterone implants in the lateral midbrain release aggressive behavior, while those in the preoptic region release copulatory movements. When both areas are simultaneously stimulated by hormones, the complete sequence of courtship behavior is more readily released. These results also support the supposition of a mixture of sexual and aggressive motivation for many courtship patterns (see Section 8.2.2).

The comparisons that are possible to date with these two experimental methods have shown that autoradiographically identified areas of concentration in the brain and those sensitive to the implantation of hormones are apparently identical. Thus, the courtship behavior of castrated ringdoves can be reactivated by testosterone implants in the hypothalamus, which also shows the greatest concentration of radioactive materials after injections. In rats and cats, similar correspondence was found. Hence, this can be accepted as proof that the control of behavior that is dependent on hormones resides in specific centers of the brain.

5.2.3.4. Determining Effects on Brain Structures

The strongest influence on the brain by sex hormones is not exerted as one might expect on adult, sexually mature animals. Instead, it is most dramatic in young animals very early in life, shortly after birth, or even earlier. This influence is characterized especially by its permanence. While the activating effect in the adult organism is always of only short duration, hormone effects in the first days of life or during embryonal development can affect behavior for the rest of the animal's life. This is especially true for the determination of sex role, i.e., whether an animal will behave primarily as a male or as a female. Hormones apparently have a determining influence on certain brain structures during this early phase, and influence them in either one or the other direction.

This permanent and later almost irreversible effect is similar to the long-term effect observed in the area of learning with respect to the memory traces involved in imprinting processes (see Section 7.4.7). Hence, one often speaks of "hormonal imprinting." The correspondence with true imprinting can be carried one step further, since the appropriate effect of the hormones is possible only during a species-specific period during ontogenetic development. This is referred to as the SENSITIVE PHASE, a term also used in imprinting terminology.

Methodologically, the determining influence of hormones can be tested in two ways: young animals can be castrated early in life, hence their own hormone production can be eliminated from the beginning, or one can supply the hormones of the opposite sex to castrated or intact animals. This can be accomplished by injections, implantations, or the transplantation of gonads from the oppostie sex. In mammals in which the embryo is connected to the mother's circulatory system, hormone treatments can be carried out with pregnant females, and the young animal can be influenced even before birth. Even "chemical castration" (see Section 5.2.2.1) is possible in the mother's body. In birds, this "prebirth" influence is achieved by submersing eggs in appropriate hormone solutions.

The results of such experiments show that early treatment with hormones will result in an almost complete and lasting change of sexual behavior in an animal. Consequently, male dogs that were castrated hormonally in the mother's body will not urinate with a raised leg but will instead do so in the squat position characteristic of females. Male rats whose testes were removed immediately following birth react like females when they later meet intact males. In females rats and guinea pigs, however, female sexual behavior can be suppressed or eliminated if the pregnant female is treated with androgens. If this treatment is continued after birth, male behavior patterns will eventually appear.

The determining effect of the sex hormones seems, however, to be limited primarily to the androgens, since similar treatments with female hormones do not show a comparable effect. Overall, one can conclude, on the basis of existing results, that the control centers for sexual behavior are still undetermined within the embryo, i.e., they are independent of the genetically determined sex. They become channeled in a specific direction about the time of birth. There seems to be a primary tendency toward female differentiation in the brain centers. In males, they can only be directed in the male direction by the specific effect of the androgens. In other words, the "hormonal sex" is determined by the presence or absence of male sex hormones during a specific developmental phase. The transitory activation of the testes, which can be observed in many species around the time of birth, probably has some kind of relationship to this task. For the expression of female characteristics, no hormonal impulses seem to be required.

In mammals, the hormonal determination of the sexual control centers is related to the temporal release of the gonadotropins by the pituitary gland. This secretion is continuous in the male sex: in females, where it is responsible for the regular ovulation periods, it appears in cycles. Experiments with rats and guinea pigs have shown that the cyclic functional type in the female can be completely suppressed by early treatment with androgens. In males, it can be artificially produced by castration immediately after birth.

Despite the seeming differences between the hormonal influences on the brains of adult and young organisms, both events seem to be closely connected. It is possible that the nature of hormonal determination consists of a sensitization of certin elements in the brain, i.e., their later sensitivity to certain hormones is increased. Hence, the presence of testosterone at the appropriate time during a critical phase of development can sensitize the corresponding brain areas for male hormones, and desensitize them against female hormones. On the other hand, the absence of testosterone increases the sensitivity toward female hormones and thus results, in both sexes, in a readiness to exhibit female sexual behavior. Just how such a hormonal sensitization is brought about is completely unknown at this time.

5.2.3.5. Production of Messenger Substances

In all instances reported so far, the hormones exert their influence directly, i.e., without any intermediate substances. Some hormones, however, work only indirectly. In this instance, the specific effect occurs some time after hormone treatment. Thus, in cats, radioactively marked estrogen concentrates a few hours after administration in certain areas of the brain. However, it disappears again without showing an immediate effect on behavior, but 4–6 days later, sexual behavior is in evidence. Estrogen apparently causes the production of some messenger substance, which, in turn, must be present in sufficient quantity before it can influence behavior. Indeed, it is now known that estrogen stimulates the neurosecretory activity of brain cells and can thus influence the pituitary gland by means of these releasing factors.

5.2.4. Hormone Release Controlled by Behavior

The preceding compilation shows in how many ways hormones can control behavior. The relationship is not, however, one-sided. Instead, the performance of certain behavior patterns, or even the perception of behaviorally important stimuli, can influence the activity of hormone glands. Again, such an effect is most readily apparent in the area of reproduction, as in the care of young and in the relationship of sexual partners to one another.

Thus, the visual, acoustic, and tactile stimuli emanating from rat pups release a substantial amount of the hormone prolactin within 30 minutes (see Section 5.2.2.4). The sucking behavior of the young exerts a strong facilitating influence here. In birds, seeing and feeling the eggs during incubation also results in the increased secretion of prolactin. The stimulating and synchronizing effect of courtship behavior (see Section 8.2.2.4) is achieved with the participation of hormones. In experiments with ringdoves, the stimuli from a courting male influence the gonads of the female via the brain and the pituitary gland, which in turn influences the behavior. At first, the pituitary gland is stimulated to secrete an increased amount of FSH (see Section 5.2.2.4). This results in an increased secretion of estrogen in the ovaries, which in turn has a facilitating influence on the nest-building behavior of the female (see Section 5.2.2.3). The performance of nest-building activities in turn stimulates the production of LH in the pituitary, which controls ovulation and the secretion of progesterone in the ovaries. This finally is responsible for the appearance of incubation behavior.

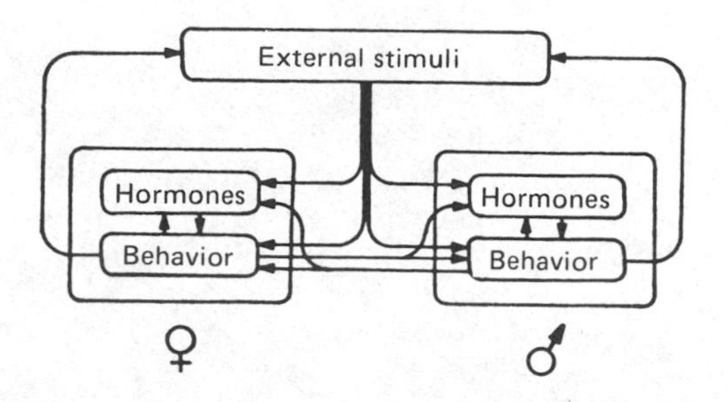

Fig. 36. Schematic representation of the reciprocal dependence of hormones and behavior. This example pertains to the reproductive behavior of ringdoves (after Lehrman 1964).

Similar influence can also occur within the same sex, especially in species that live in groups during the reproductive period (see Section 8.4.1.3). This is true for the budgerigar, where the song of the male not only influences the behavior and physiology of the female (see Section 8.2.2.4) but also stimulates the growth of the testes in other males, and hence their production of testosterone. It is very likely that the temporal synchronization of incubation observed in the breeding colonies of many birds species probably depends on such mutual stimulation.

These few examples show that long-term behavior sequences and changes can be controlled by the reciprocal effects of hormones and behavior patterns. A behavior pattern facilitates the secretion of a hormone, which in turn influences the release of an additional hormone. An important link in this reciprocal relationship is those cells in the CNS that are capable of transforming hormonal and nerve impulses, respectively, into another form of information. This includes certain cells in hormone-sensitive areas of the brain (see Section 5.2.3.3) that transmit hormonally received information to the control centers in the CNS. In the reverse order, neurohormonally active brain cells secrete releasing factors in response to nerve impulses that affect the activity of hormone glands (see Section 5.2.3.5).

6

Ontogeny of Behavior

ONTOGENY is the development of a living being from the time the ovum is fertilized until the organism dies. In ethology the development of behavior in the young is of special interest, i.e., the period from birth to the attainment of sexual maturity. During this time the greatest changes in behavior take place, since

- ☐ the complete behavior of a species develops gradually during this period and attains complete expression.
- ☐ during early childhood development phylogenetic rudiments (see Section 10.5) may be present that later are lost again.
- ☐ the behavior of young animals shows a number of special adaptations that are suited for their own particular needs and that later disappear. This is most strongly expressed in species whose development includes a larval stage.[1]

Hence, behavioral ontogeny offers especially favorable conditions for the investigation of the various kinds of developmental processes. These encompass two areas: the description of the development of behavior as such, and the concurrent maturational and learning processes along with the analysis of the internal and external factors that influence this development. More recently, prenatal development of behavior, e.g., learning processes taking place before birth (see Section 7.2), has been investigated to a larger extent.

6.1. Inborn or Acquired?

6.1.1. Sources of Information

The relative contributions of inheritance and environment to the development of behavior has been discussed since the early days of ethology. These discussions were also of interest outside ethological circles because of the possible significance of these questions for human behavior. Human and social

[1] In biology a LARVAL STAGE refers to a free-living stage early in life that is substantially different from the life-style of the adult, sexually mature animal. Such adaptations are characterized by the presence of special organs (so-called LARVAL ORGANS).

psychologists were influenced by these issues, and they in turn influenced and contributed to the discussion. In some instances the importance of inheritance was certainly overemphasized, while others undervalued or even denied it. This denial of the importance of inheritance—when not ideologically rooted—was frequently based on the assumption that genetically and environmentally determined components of behavior are so interrelated during development that it is impossible to separate them, and secondly, that the distinction would be heuristically without value.

Here we have to state a fundamental premise: The behavior of each animal is adapted to its environment. This is true for movement patterns in search of food or in the avoidance of enemies—and it also applies to social behavior. Each part of an animal's behavioral inventory is rather perfectly attuned to the requirements of its living and nonliving environment. Many of the more recent ethoecological investigations have shown how such adaptations can extend to the social structure or the learning program of a species.

Each adaptation presupposes corresponding information. In principle, these can enter an organism in two different ways: by means of the genome and via the sense organs. Information can be stored in the genome, which represents a "species memory" that is passed on to each successive generation and can be stored in the "individual memory" of an animal. This is a fundamental difference. Hence, the use of two separate concepts, which denote clearly the respective source of the information, seem to be justified. If it is derived from the genome, the behavior is referred to as inborn or innate; if it is based on individual experience, we call it acquired or learned. In emphasizing the source of the adaptations, the terms INHERITED ADAPTATIONS and ACQUIRED ADAPTATIONS are sometimes used.

No ethological concept is as controversial in its application as are the terms INBORN and INNATE. There were and there still exist many misunderstandings. It was frequently imputed that these terms were meant to indicate an independence of the behavior from the environment that would make it unresponsive to external influences. With respect to the term ENVIRONMENT, however, one needs to make two fundamental distinctions: the general environmental conditions to which an organism is exposed during its development, and the specific environmental factors to which the behavior pattern whose development is being studied is exposed during the course of its development. In the former sense there is no characteristic of an organism, including behavioral ones, that can develop without being influenced by the environment. Even the genes transmit their effects only by continuous interaction with their environment. The same is true for tissues, organs, and all other holistic characterisitics whose development is continually influenced by conditions internal and external to the organism.

Matters are somewhat different with specific individual adaptations. A prey-catching behavior pattern can manifest itself in a predatory insect immediately after it hatches from its pupal case. The behavior is complete and appropriate and therefore does not require specific individual experience with any prey. In this case one speaks of an INNATE BEHAVIOR PATTERN. This concept simply means that the information for the behavior was stored in the genome of the species during the course of its evolution. Hence, it does not need to be acquired anew by interaction of the environmental conditions with those of the behavior—in this case the attributes of the prey. Even here, however, the environment is involved. In most instances even an innate behavior is gradually perfected by experience during the course of an animal's life.

The concept of *innate* is hence not meant in any way to state an exclusive hypothesis. It does *not* say that the environment has no part in the formation of a behavior characteristic. Quite the opposite is the case: Inherited is always the norm of a reaction, an "offer," as it were, to the environment. Within this range of expression various environmental influences decide how the information emanating from the gene will ultimately be expressed in the individual case. This holds for all other characteristics of an organism in exactly the same way. It is true especially with learned behavior, where the constant reciprocal effects of environment and inheritance are clearly recognizable. One example may illustrate this: Some learning processes can take place only during a very specific age, which varies from one species to the next, but which is quite fixed within a species (see Sections 7.2 and 7.4.7.1). Here inheritance determines at which age something *can* be learned. However, within these temporal constraints the available learning possibilities decide when, in the individual instance, something will be learned, i.e., whether the animal will store the specific experience in its memory in the beginning, at the end, or in the middle of the learning phase.

6.1.2. Recognition of Inborn Behavior Patterns

The question of whether a behavior is inborn can only be decided by an experiment. The best evidence comes from behavioral genetic investigations. However, these are only possible in relatively few species due to methodological considerations (see Chapter 9).

A substitute method is available by raising young animals "without experience", i.e., subjecting them to a period during which normally impinging environmental stimulation is withheld. Animals that have grown up under these conditions are called Kaspar Hausers (K.-H. for short).[2]

A complete withholding of experience (a so-called K.-H. of the first order) cannot be achieved, since there is always a remnant of possible experience, e.g., with the animal's own body, that cannot be excluded even under the

[2] The term KASPAR HAUSER refers to a foundling of this name who appeared in 1828 in the city of Nuremberg, Germany, and attracted much attention. There is substantial literature surrounding his fate, although his origin could never be satisfactorily explained. According to his own statements, he had always lived in a dark room. His mental development remained limited throughout his life. From this disruption of normal development in his youth, aberrations from normal behavior remained that are parallel to those that ethologists find in experimentally produced Kaspar Hauser animals. The term DEPRIVATION EXPERIMENT is also used in the English literature for this kind of experiment.—TRANS.

most rigorous conditions. Nor would the results of such an experiment be very valid, since the disruptions of the developmental sequence would be such that a comparison with the normal behavior of the species would no longer be possible. Partial K.-H.s present fewer problems, since one can withhold those specific stimuli that play a role in the particular functional system under investigation. For example, if one wants to know if a species has an innate preference for certain foods, then it is only necessary to withhold all experience with the normal food of the species. Later one can test the animal so raised to see if there is a spontaneous preference for that food. The same applies to the recognition of enemies or conspecifics. In these instances one speaks of a *prey K.-H.*, an *enemy K.-H.*, or a *social K.-H.* Finally, it is possible to exclude individual sense organs, such as by covering the eyes, or making the animal artificially deaf, and the development of certain behavior patterns can be studied under conditions where these stimuli are absent. Investigations of *acoustic K.-H.s* have contributed valuable insights about the contribution of inheritance to the vocalizations of animals.

All K.-H. or deprivation experiments share one disadvantage: One can never completely exclude the possibility that such an artificial, more or less isolated raising of animals, especially social animals, will not be in some way detrimental to the animal. With reference to sense organs, there may be a degeneration of the retina in dark-reared animals, which in turn may affect the development of behavior quite aside from any effects of the withholding of specific experiences. Hence, the interpretation of results demands extreme caution. In principle, valid statements about the experimental results are possible in only one of the two possible outcomes: If a behavioral characteristic, e.g., a movement pattern or an object preference, develops normally in a K.-H., then one can say with some justification that it is innate. If, on the other hand, the result is negative or the behavior develops abnormally, the conclusion that it is learned does not automatically follow. Such a result may point in that direction, but the absence of the behavior could also be the result of some disruptive influence during development.

In many instances it is not possible to conduct behavioral genetic or deprivation experiments. In order to obtain some indication in a specific case as to whether one is dealing with innate behavior, a number of criteria have been established that allow some tentative conclusions in this respect.

Genetic programming of behavior may be involved if

- ☐ a behavior is characterized by great constancy or stereotypy in form, is found to be the same in all members of a species, and cannot be influenced in its performance by external stimuli.
- ☐ a complex behavior occurs in its complete form the first time it is performed.
- ☐ a behavior appears before the morphological structures that are involved are fully developed.

The first criterion is the weakest in terms of validity. While it is true that many innate behavior patterns, e.g., courtship or threat behavior, are highly stereotyped, learned movement coordinations can also attain a high degree of form constancy during ontogeny (*ontogenetic ritualization,* see Section 10.6). The equation of innateness and form constancy has been made by some, but it is not warranted. The general distribution of a characteristic within a species is also not conclusive evidence for innateness, since a characteristic may be the result of parallel learning processes that took place in identical environmental conditions. This is especially true with obligatory learning processes (see Section 7.3).

The two other criteria permit more definite conclusions: When a young spider builds a complete net for the first time in its life just like an animal that has built one hundreds of times before, and if the complex behavior patterns that are required and the proper order in which they must occur take place, then genetic preprogramming is the only possible conclusion. The same is true when movements are performed with organs that have not yet attained their final form and size, so that neither can the "goal" of the behavior be achieved nor was there an opportunity to practice the behavior. Thus, young goslings show typical fighting movements with their wings at an age when they are still so short that they cannot even reach their opponent. Even more obvious are those instances in which the behavior does not yet take place before the development of the appropriate morphological characteristics, but where the impulse patterns from the CNS are already present. The recording of action potentials (see Section 5.1.1.3) was possible in a field cricket before the emergence from the the last instar, i.e., before the sexually mature animal emerged. The complete programs for all song types (see Section 5.1.2.1) could be demonstrated, although the animal could not have practiced them in its larval stages because its wings had not hardened and were not completely developed.

Finally, it must be emphasized that all three criteria are only indications, albeit very obvious ones, but they do not constitute absolute proof of the innateness of behavior. Hence, this term should be used with extreme caution.

One criterion, used in medicine to identify hereditary diseases and other characteristics, is completely inappropriate for use in ethology: the presence or absence of a characteristic at birth. There is learning before birth (see Section 7.2), based on which the behavior of the newborn can already contain learned components, whereas innate behavior components may appear only gradually as a result of maturation (see Section 6.2). Behavior patterns that depend on sex hormones do not appear until a certain age is reached in long-lived vertebrates, and in some insects they appear only after many years (in the cicada up to 17 years), although these behaviors are partially or largely preprogrammed, as can be assumed in insects.

6.1.3. What Is Inborn?

With respect to the *kind* of innate components of behavior there are two possibilities: The performance of a movement can be preprogrammed genetically, or the knowledge of releasing and directing stimuli can be inborn. It has been found that the former is more prevalent.

European polecats *(Putorius putorius)* raised without prior experience in hunting rats are able to chase them, throw them off balance, and grab and shake them (see Section 2.2). They show the species-specific prey-catching behavior patterns, but grab any part of the body they happen to catch. They learn only gradually which part of the body should be bitten for more efficiency in killing the prey: the neck. The proper orientation of the killing bite must be learned. Young ducks and geese innately follow a moving object shortly after hatching, but they must learn *whom* to follow. In this case, the knowledge of the releasing stimuli needs to be acquired. Behavior patterns that are genetically fixed, for which the knowledge of the releasing stimuli must be learned, have evolved without reference to the proper object. Elsewhere (see Section 3.9.1) we have discussed inherited movement coordinations, where the indications are that a contraction patterns of muscles, and the spatiotemporal coordination of the individual movements are genetically programmed.

It is much rarer to find an innate coupling between the releasing stimulus and the released behavior pattern. In such instances an animal reacts in the appropriate manner to a stimulus pattern without ever having encountered it before, i.e., without having had the opportunity to learn about it. A number of well-investigated examples are available to illustrate this: Herring gull chicks direct their first pecking reaction shortly after hatching to a red spot on a yellow background as much as do experienced animals that are several days old (see Section 1.3). Pied flycatchers, which were hand-raised and had no opportunity to learn about their predators, still show the typical alarm reaction, the so-called mobbing response, against a living owl. In some cases the reaction is as strong as in "experienced" birds. Some North American snakes show highly specific food preferences at birth that even vary in different parts of their range. The animals "know" what to do when they encounter a certain stimulus for the first time. The knowledge of the red underside of a stickleback male, which constitutes an aggressive signal for other males and a sexual signal for females, is innate. Animals raised in isolation also respond to it in

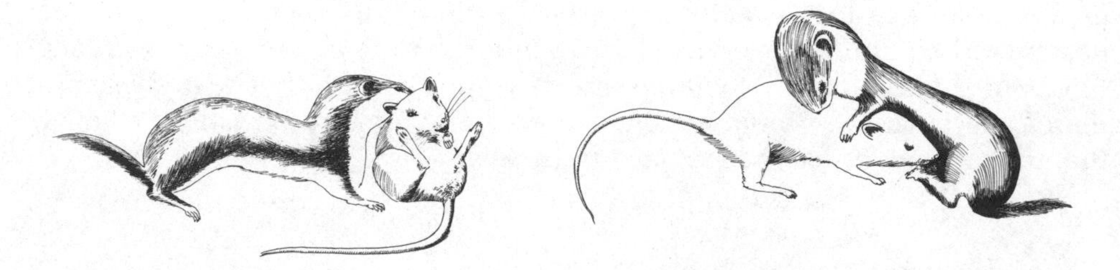

Fig. 37. Prey-catching behavior in the weasel. An experienced animal kills the mouse by a neck bite (left); an inexperienced one, on the other hand, bites into the back of the prey (after Eibl-Eibesfeldt 1963).

an appropriate fashion. With respect to this cue, a "natural experiment" involving females was recently described: In western North America there is a population of the three-spined stickleback in which the males are of uniformly black coloration. This characteristic is probably an adaptation that evolved as protection against a predator only found in this region. If one presents the females of this population with one black and one red-bellied male, an unexpected result is obtained: the females prefer the red-bellied males 5:1 over the black ones. Since the adaptation to the predator probably evolved during the last glacial period, when this population became isolated from other stickleback populations, the age of the black coloration in the male can be estimated to be 6000–8000 years. The females, then, still react preferentially to the releaser in the males that was lost thousands of years ago in adaptation to new environmental conditions. A better example for a genetically determined reaction to a specific stimulus can hardly be imagined.

In the case of an innate recognition of a signal, the genome determines the filtering of the stimuli to be responded to. This is referred to as an INNATE RELEASING MECHANISM (IRM) (see Section 3.2). It is in contrast to the ACQUIRED RELEASING MECHANISM (ARM) for which the characteristics of the releasing stimuli must be learned.

As a criterion for the recognition of an ARM the relative simplicity or lack of component cues is often cited. An innate reaction can always be expected when an animal reacts within a particular functional system only to very *specific* components or cues of an object, especially if one knows from the behavior in other functional systems that the animal can well perceive other cues as well. Examples of this are the courtship flight of the grayling butterfly and the attack behavior of the male stickleback (see Section 3.1.2). However, since there are also very complicated IRMs, this characterization has its limits.

As in the realm of movement patterns, the rule seems to hold also for sensory perception that inheritance plays a large role especially in lower animals. During the phylogenetic development of animals toward higher forms, individual experience seems to become increasingly important. Hence, pure IRMs are known especially in invertebrates, whereas in vertebrates the innate knowledge is usually limited to a few general characteristics of an object. These provide, as it were, the framework that can be filled in by additional learning. For example, newborn ungulates search for the nipples between the fore and hind legs of their mothers; they learn the exact location by experience. A very similar narrowing of object knowledge takes place with animals that follow their mother during the sensitive imprinting period (see Section 7.4.7.1).

In this example, which is a rather widespread phenomenon in vertebrates, one speaks of an INNATE RELEASING MECHANISM MODIFIED BY EXPERIENCE (IRME). Here we deal with a releasing mechanism that is based on an innate tendency to respond but needs to be completed by learning during the development of the individual animal.

Another important question relates to the *extent* of innate behavior components. It seems that in no animal species is courtship behavior or threat behavior inborn as such. Usually, only components of these behavior complexes are preprogrammed in the genome. In general the rule applies (with exceptions) that in lower animals long or complex movement coordinations can be inborn, whereas in more highly evolved forms—especially in vertebrates but also in some invertebrates, e.g., cephalopods (and in some insects like sawflies and wasps)—innate components are fewer and the sequences shorter. The mutual interaction between genome and environment can be seen down to the smallest subunits of behavior.[3]

The extent of innate behavior patterns is to some degree correlated with the LEVEL OF INTEGRATION at which the genetic programming exists. When short, single movements are inborn, which are combined into a meaningful sequence only by experience, then integration exists on a lower hierarchical level (see Section 4.3) than if longer movement coordinations take place independently of prior experience. If one raises rats without giving them the opportunity to handle any objects, by feeding them a liquid diet and by keeping them on a grid through which their feces drop, and if one furthermore amputates their tails, they are still able upon first presentation of nesting material to perform the individual, species-typical behavior patterns associated with nest building (splitting, collecting, and stacking materials, forming a nest cup). However, they do not produce a real nest at first, because they must first learn to integrate these individual behavior patterns into the proper order. The same was observed with nest building in inexperienced canaries.

The level of integration at which one characteristic behavior of the European red squirrel, burying its food, is organized is much higher. Animals that were raised simply on soft food in wire cages not only perform, upon first presentation of a nut at the appropriate age, the specific movements (scratching in dirt with the forelegs, depositing the nut, tamping it down with the snout, covering it with dirt, and tamping down the dirt with the forelegs) but they also perform these actions in the same sequence as experienced conspecifics. They perform this behavior even in a bare cage or under a blanket, where they can neither scratch in dirt nor cover the nut. Hence, the entire movement sequence appears rather inappropriate. This is a good example for the stereotyped nature of such centrally programmed movement coordinations.

6.2. Maturation of Behavior Patterns

In correspondence with the two sources of information discussed earlier, the ontogeny of behavior is made up of two processes: The transformation of information that is passed on by the genome, and the acquisition of new

[3] The term INSTINCT–TRAINING INTERLOCKING was used for some time for this interaction between innate components and learning processes. However, since it is difficult to limit or define instinctive behavior and training, respectively (see Sections 4.3 and 7.4.3), this combination of terms is only rarely used today.

information through the sense organs. The first process is referred to as *maturation*, the second as *learning*. During ontogenetic development both processes are closely intertwined. It is possible to separate them experimentally, and to identify them and to learn about the extent of the respective contributions of each process.

Since most learning processes are not limited to the juvenile period, it would not be appropriate to discuss them solely in connection with the ontogenesis of behavior. Hence, we will discuss here only maturational processes.

MATURATION in ethological usage refers to the full development or perfection of a behavior pattern without practice. A behavior matures if it improves during ontogenesis, even if there is no opportunity to perform the behavior. Maturation processes can be demonstrated if one prevents young animals from performing a behavior during the period in which it normally unfolds, and by comparing control animals that were normally raised with the experimental animals. For example, pigeons that were kept in small cages in which they were unable to raise their wings are capable of flight just like controls that could exercise their wings and flutter about before their first actual flight. Both groups did this at the same age. Tadpoles that were kept in a state of continuous immobilization by a narcotic substance swam as well as control animals when the drug was removed. Siberian hamster females build incomplete nests that are too small when they give birth for the first time at about 80 days of age. Females raised as Kaspar Hausers under identical conditions, which are at least three months old when they have their first litter, build nests that cannot be distinguished from those of normally raised conspecifics.

Even an aiming mechanism, i.e., an ability to orient, can mature, as was shown in experiments that investigated the food-pecking response in chicks: If chicks are presented with a nail embedded in clay, they will peck at it as they would at food. On the first day of life their pecks, which are seen in the clay, often miss and are scattered about the shiny nail head. Their aim improves during the following days as can be seen by the pattern of the pecks—most of which now hit the target. This improvement suggests a learning process, especially since the chicks also peck at seeds between experiments where they could be rewarded for improving their aim by trial and success. This same improvement in pecking accuracy, however, is also obtained by chicks that from the first day were fitted with prism goggles that deflected their visual field by 7° to the right or left. Their pecks then fell either to the right or left of the target when the chicks aimed at it, but they never actually hit it. The chicks in this group were given their food in a bowl so that there was never a difference in reward whether they actually hit or missed a seed. In spite of this their pecks became more clustered, although displaced to the sides of the target, just as in the control group—a clear indication that no learning by reward process was involved. Hence, the conclusion that a maturational process that functioned independent of learning was involved.

These examples show that maturation and learning processes appear very

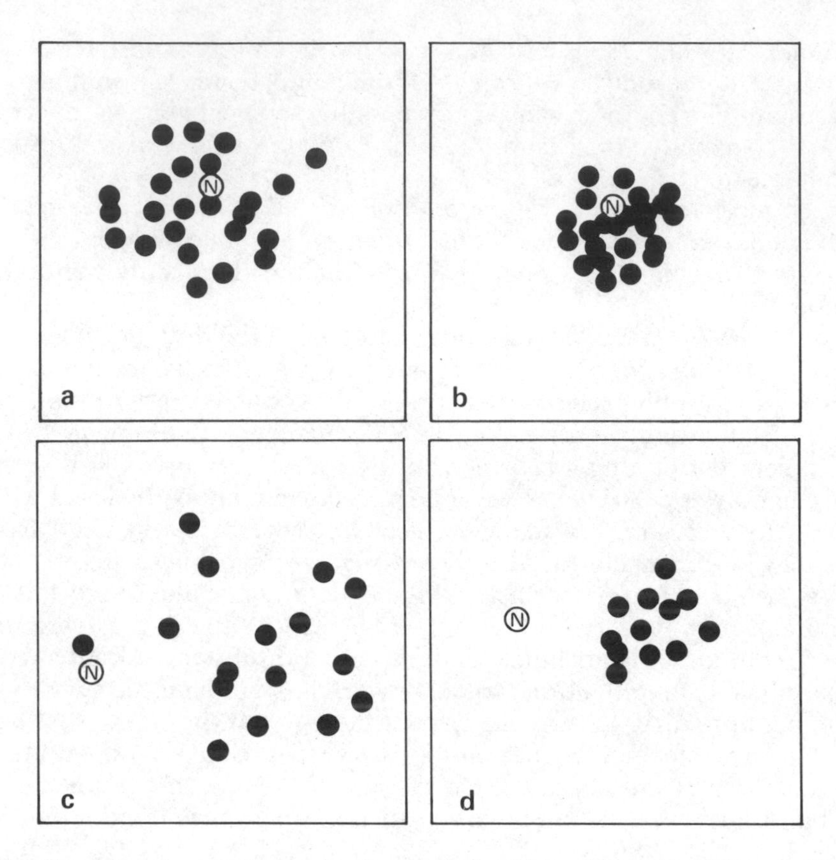

Fig. 38. Maturation of an aiming mechanism. Pecks by chicks without (a and b) and with (c and d) prism goggles (N = nailhead; additional explanation in text) (adapted from Hess 1956).

similar, although the underlying mechanisms are different. Maturation depends neither on a training of muscles nor on an improvement of the results of maturation. Instead, it must depend on developmental processes in the central nervous system. Maturational processes can also give us an indication about how genetic information can be translated into behavior patterns. Of course, no behavior pattern is directly controlled by genes. Instead, information contained in the genome leads at first to a gradual, i.e., maturational, development of a "behavior program" in the central nervous system. Only then can the CNS determine the course of a behavior via appropriate impulse patterns whose characteristics have already been discussed (see Section 5.1.2.4). These central impulse patterns—and not the genetic programs—can in many instances be improved by the interaction with individual experience, i.e., additional adaptations to environmental contingencies are thus achieved.

7

Learning

In ethology the concept of LEARNING encompasses all processes that lead to an *individual* adaptation of behavior to particular environmental conditions. Learning requires the acquisition of information with the sense organs, its storage in memory, and recall when needed. Learning processes are known in all classes of animals, from single-celled[1] animals to vertebrates. They are recognized in general by the changes in behavior that occur in a specific stimulus situation as a result of prior exposure to this situation.

With respect to learning behavior in animals, four general questions can be asked. We will attempt to answer them in the following sections:

- ☐ Why is there learning?
- ☐ When does learning take place?
- ☐ What is being learned?
- ☐ How is learning accomplished?

7.1. The Biological Significance of Individually Acquired Information

If the biological advantages and disadvantages of inherited and acquired information are to be evaluated against one another, it is necessary to distinguish between two situations: one in which both occur "without competition"—hence, where only one kind of information transmission is possible, and another in which theoretically both can be realized but where they compete with one another during phylogenetic development.

An example for the first condition includes the individual knowledge of family and group members (see Section 8.4.4.2), which, because of the lack of predictability, is only possible by learning processes. As to inheritance, examples are species in which one generation is dead before the next one is born, and where all characteristics such as escape and avoidance reactions, certain building techniques, or the recognition of a conspecific sexual partner cannot be acquired by an individual animal through trial and error but *only* by inheritance through the genome. Hence, many insects and other invertebrates literally grow up as social Kaspar Hausers.

However, in most instances, information can be acquired both ways. Since learning processes are so widespread, and increase in importance with the

[1] The results from single-celled animals need a critical reexamination.

level of evolutionary development—in other words, in the individual case, learning processes are phylogenetically younger than genetically acquired information—then they ought to have the same advantage over the latter as prevailed during the course of evolution.

This advantage seems to be the same for all learning processes: It consists of the greater adaptability of acquired information. The environment of a species changes continuously. This is true for the physical environment such as climatic changes, as well as for the living environment, e.g., the appearance or disappearance of predators or food. These changes also take place within a species, e.g., changes in appearance that occur as a result of mutations. Each of these changes can be adjusted to within one generation by learning processes. Adaptations within the framework of a genetic program are only possible on the basis of mutations whose temporal appearance is just as "arbitrary" as is their direction (see Chapter 10). We cite here only the reaction of black stickleback females to red-bellied males (see Section 6.1.3), a classical example for the conservatism of an innate releasing mechanism. These differences are especially obvious in the selection of food: When a species specializes on one or several kinds of food and recognizes them innately, it is doomed to extinction when this food disappears. If, however, the species has an appropriate learning capacity, an adaptive change is possible at any time.

Hence, it should not be surprising that the overall phylogenetic trend is in the direction of an increased development of learning capacities. This replaces a "closed" developmental program in which genetic programming predominates. An "open" program, on the other hand, leaves enough leeway for individual experience, and hence constitutes a true adaptive advantage.

The selection pressure in the direction of a quick adaptation differs, however, in various functional systems. This pressure is greatest with respect to ecological factors, and is less in various social areas because changes, which require an adaptation (e.g., in the appearance of species members that need to be known), are usually much slower here than changes in the extraspecies environment. Hence, social behavior, such as courtship or threat behavior patterns, shows the largest component of genetic programming. They offer the best indications of phylogenetic relationships (see Section 10.2).

7.2. Sensitive Phases in Learning

Certain learning processes, imprinting before all others (see Section 7.4.7), are not available during the entire life but are limited to specific ages. These are referred to as SENSITIVE PHASES,[2] a concept that is also used in develop-

[2] The term CRITICAL PERIOD, which was used especially in the early investigations on the imprinting phenomenon, is misleading, and its use should be avoided. The term PHASE is used here in the sense that it is nonrecurrent. However, this stage of an animal's life is by no means "critical"; rather, it is an optimal and positive phase, especially for this kind of learning process. A critical situation does not occur until after the termination of this phase in the event that the required learning experiences did not take place at their normal time. Under natural conditions, this does not occur.

mental physiology, from which it was adopted. Usually this period occurs early in development, i.e., during the first few days or weeks of life, and it is quite short with reference to the animal's life-span.

The biological significance of early phases of maximal learning ability is easily recognized: Early in its youth, the animal is especially close to species members (parents, siblings, other family and group members). Therefore, it can acquire knowledge and experiences that are needed later in life more readily than after the dissolution of the family group. Hence, it is useful for an individual to be especially receptive for these kinds of impressions and experiences. An increased sensitivity during this time would certainly be of high adaptive value.

Little is known thus far about the physiological basis of this exclusive or maximal learning ability during limited periods in an animal's life—in spite of many investigations and equally numerous speculations. Most likely, developmental processes in the CNS play an important role.

One developmental phase, with respect to such learning processes, that has recently attracted special interest among ethologists occurs before "birth." In many bird species, it is known that well before hatching, still in the egg, the animal can acquire information from the outside. It can even enter into communication with its mother and siblings. Thus, quail embryos exchange calls "from egg to egg" that result in the synchronization of hatching. Embryos that lag in their development are positively influenced by the calls of their advanced siblings, while the somewhat weaker calls of the late developers seem to have a retarding effect on their more advanced siblings. Such synchronization is extremely appropriate since it enables all the chicks to leave the nest with the mother at the same time.

In mallard ducks there is a lively acoustic exchange between the embryo that is ready to hatch and the incubating female. This can also be done artificially by playing back recorded sounds from one to the other. In all examples cited thus far, it is not known whether learning already occurs before hatching. However, in chick embryos this is known with certainty: If certain sounds are played into the incubator during the incubation period, these chicks prefer the familiar sound to unfamiliar ones after hatching. Under natural conditions we know of learning before birth in the guillemot, a seabird that nests on narrow ledges on rocky coasts and islands. The adults respond to the calls, from the egg, of their chick, which thus learns to recognize its parents and can eventually distinguish them from neighboring birds. Thus, by the time the chick hatches, there already exists an acoustic bond with its parents. This is a necessary precondition for the successful raising of young in these dense breeding colonies where pairs breed together in very close proximity.

In mammals the embryo is much more isolated from the environment. Prebirth learning processes hence would not be expected to the same extent, but they possibly occur in a more limited sense. In primates there are indications that the rhythm of the mother's heartbeat is already perceived by the embryo, and that this signal may play a role later in the mother–child bond.

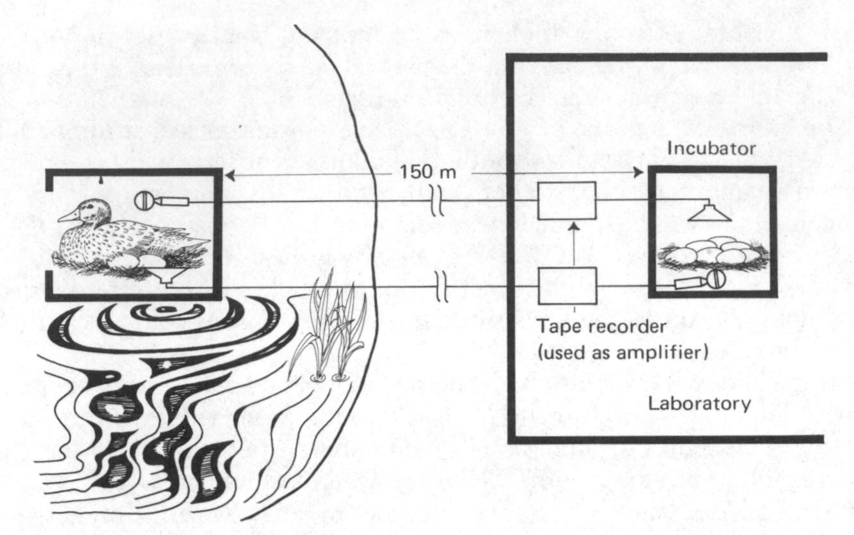

Fig. 39. Experimental design for the investigation of auditory communication between an incubating mallard female and her ducklings, which are about to hatch (after Hess 1975).

7.3. Learning Predispositions

The extent of the learning capacities of a species corresponds roughly to two criteria: (1) the phylogenetic level of development, and (2) the specific environmental conditions in which it lives.

In general we can observe an increase in learning capacities as we go up the phylogenetic scale. The biological significance of this has been discussed in Section 7.1. It is achieved by an increasing centralization and perfection of the nervous system, which has attained its highest development in the brains of arthropods and vertebrates.

Within this general extent of learning abilities, there seem to be differences that are not correlated with the level of phylogenetic development, since they exist even between closely related species and even different races of the same species. Such special adaptations indicate a close correlation to the life-style of the particular species. One of the most frequently cited examples is the recognition of young in gulls: The parents of ground-nesting species learn to recognize the young individually within a few days after hatching. As soon as the young are able to leave the nest and run about in the colony, the parents are able to recognize them. If one attempts to exchange the young toward the end of the first week of life, they are no longer accepted. The behavior of the kittiwake gull is quite different. In this species the parents never recognize their young individually, and they will accept strange young when these are already four weeks old. This different behavior can be readily explained with reference to the breeding biology of the species. In contrast to other gulls, this species nests on even the smallest ledges on cliffs that are just large enough to offer space for two young. As an adaptation to this nesting strategy, young kittiwakes do not leave their nest until they fledge. Hence, the parents never need to search for and identify their young at another location. It is enough if they recognize the location of their nesting place.

Intraspecific differences in learning capacities were discovered in the honeybee. Learning experiments have shown that members of the Krainer race (*Apis meliffica carnica*) are better able to learn visual landmarks in search of a feeding place than are bees of the Italian race (*A. m. ligustica*) under identical experimental conditions. This learning predisposition is biologically meaningful: Italian bees are able to use the sun as a reference point in orientation more frequently, owing to the better climatic conditions in their home range. Krainer bees, which live in an area of changing weather conditions, must depend more heavily on visual landmarks, in addition to the sun, as an aid in orientation.

Similar intraspecific differences in honeybees also exist in the ability to learn odor cues.

These examples, and many others, show two things: (1) There seems to exist in each species or subspecies a strong selection pressure (see Chapter 10) to develop a special learning ability in areas important to them. The fact that even subspecies differ in learning abilities indicates the speed with which

Fig. 40. Kittiwake gull with young.

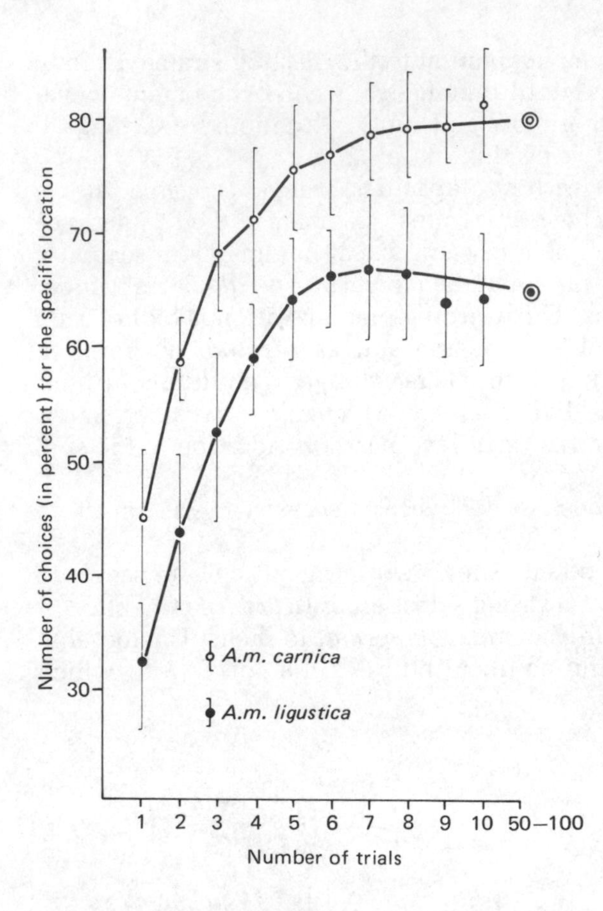

Fig. 41. Different learning curves in two races of the honeybee: The Krainer race is better in learning a task that involves visual cues ("The western point of a black star indicates a food goal") than the Italian race. The points in circles at the end of the curves show the combined values of trials 50–100 (after Lauer and Lindauer 1971).

these specific learning dispositions can evolve. (2) These learning predispositions enable us to recognize the interaction between the genome and the environment (see Section 6.1). For each species and subspecies there exists an innately proscribed range of learning ability; hence, the genome determines the limits of learning capacities.

Another difference with respect to the range of learning exists with reference to obligatory and facultative learning processes: There are learning processes that are absolutely essential for the survival of a species—and others that are possible, but not necessarily essential. In the first category are all imprinting processes (see Section 7.4.7), and learning processes in the functional systems of predator avoidance and acquisition of food. The second category encompasses individual recognition of conspecifics and learning in curiosity and play (see Section 7.4.4). Facultative learning processes are biologically not less important than obligatory ones. However, they possess more freedom of expression from inherited limitations. Hence, they can adapt more specifically to immediate environmental situations, and thus show great individual differences. However, a sharp dividing line between these two categories cannot be drawn. Laboratory and circus experiments show that the facultative learning capacity of many species is larger than its normal range of expression under natural conditions.

7.4. Learning Processes

The abundance of environmental influences that have an effect on an animal are so great, and the manner in which they influence its behavior are so varied, that it is difficult to separate learning processes from other environmentally dependent behavior changes, so as to order and classify them in some meaningful way.

One example will illustrate these problems: Rats that have been shocked early in life later learn better than those that have not. Is this learning? Did electric shocks produce a learning process, and which one? If not, how and where does later improvement in learning fit in? These problems cannot be discussed here. They were mentioned to show that the concept of learning can be used in various ways within ethology. Furthermore, its use, in contrast to the way it is used in human psychology and in everyday language, may also be different.

For all of these reasons, the number of categories of learning found in animals is quite large. The following discussion includes the most frequently used learning categories in the ethological literature:

- ☐ Habituation
- ☐ Classical conditioning
- ☐ Instrumental conditioning
- ☐ Play behavior
- ☐ Imitation
- ☐ Insight learning
- ☐ Imprinting

It must be emphasized that this classification might be just as plausible according to other criteria. Like any other conceptual systems, it by no means includes *all* possible learning processes.

7.4.1. Habituation

The ability of animals to "get used to" repeated stimuli that are associated with neither positive nor negative consequences, and to which they no longer respond, has variously been called HABITUATION, STIMULUS HABITUATION, or STIMULUS-SPECIFIC FATIGUE. These characteristics have already been discussed elsewhere (see Section 2.3). Habituation is a "negative"learning process, since an already present readiness to respond is extinguished. In some ways it is the opposite of classical conditioning, because a stimulus that initially releases a behavior now becomes a neutral stimulus.

7.4.2. Classical Conditioning

In CLASSICAL CONDITIONING a previously neutral stimulus releases a response as a result of positive reinforcement (reward)—it becomes a conditioned stimulus. The reaction so elicited is called a conditioned reflex in the sense that it depends on experience. Its characteristics were already discussed in another context (see Section 2.1).

Conditioned reflexes cannot be maintained indefinitely. The conditioned reflex can continuously trigger a reaction only if the stimulus that originally released the reaction through an unconditioned reflex (see Section 2.1) is periodically presented. Thus, the salivation reaction in the dog disappears in response to a light flash if the connection is not occasionally "strengthened" by the presentation of actual food stimuli. This loss of responsiveness is called extinction.

7.4.3. Instrumental Conditioning

INSTRUMENTAL OF OPERANT CONDITIONING is distinguished from classical conditioning in that no new stimulus becomes associated with an already present reaction, but instead a new movement is associated with the decrease of a need (e.g., reduction of hunger or thirst). Such movements, e.g., pecking at a disk or the pressing of a lever, must first occur spontaneously (see Section 2.5). If such behavior is repeatedly rewarded, e.g., by a seed, then the animal forms an association that leads to an increased performance of the behavior under the specific drive conditions, e.g., hunger. In this way the movement has now experimentally become an appetitive behavior in this functional system (see Section 2.4).

Operant conditioning is carried out primarily in mazes and Skinner boxes. Mazes consist of a system of runs in which only one correct choice at each choice point leads to the goal, while all others end in a blind cul-de-sac. At the goal the various right and left turns, which are made at first randomly by the experimental animal, are reinforced by food, for example. An important criterion for learning is the number of trials an animal requires until it reaches the goal without errors, i.e., entering blind alleys. The difficulty of the maze depends upon the number of choice points. The simplest maze is a T-maze, which requires only one choice to be made. The Skinner box, so named after its inventor, the American psychologist B. F. Skinner, consists of a box that contains the experimental animal, a manipulandum (e.g., a disk that can be pecked by a pigeon, or a bar that can be pressed by a mouse or rat), and a reinforcement mechanism (e.g. a food dispenser).

Rats and mice have been most frequently used in operant conditioning experiments. Because of their lifelong curiosity behavior, they guarantee the occurrence of the first spontaneous emission of the behavior to be conditioned in a relatively short time.

Operant conditioning was for a long time the main investigatory tool of behaviorist psychologists. This school of psychology led to many new insights

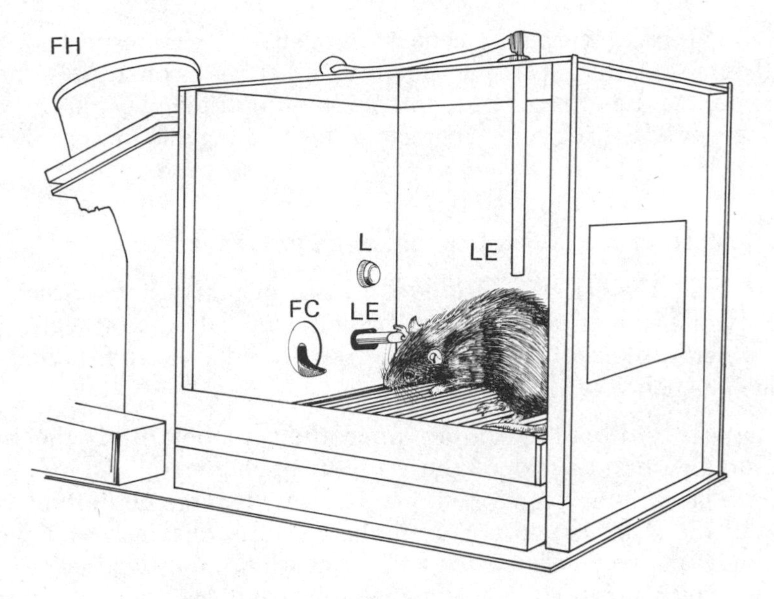

Fig. 42. Skinner box for operant conditioning of rats. LE = lever; FC = food cup; FH = food hopper, from which a food pellet is released when the rat touches the lever. By adding a lamp (L) the animal can be trained to touch the lever in response to another stimulus, e.g., the lighting up of the lamp.

into the learning abilities and behavior of animals. On the other hand, it was limited to relatively few domesticated species, and led to many unjustified generalizations and wrong hypotheses, because relatively "unnatural" questions were asked during its investigations. In recent times, more species have been investigated, and the experimental settings have been considerably expanded.

Learning processes, seen in instrumental and classical conditioning, occur not only in laboratory experiments and experimental cages. Free-ranging animals must also "try out" things, e.g., while in search of food or while building nests and other structures. Here some behaviors may prove to be more suitable than others, stimuli that indicate a specific situation are learned, e.g., a nesting site and those that contribute to the release and orientation of the appropriate behavior patterns. This learning process, which leads generally to less rigid associations than in instrumental conditioning, and in which both forms of conditioning can occur in combination, are usually referred to as trial-and-error or object learning.

7.4.4. Play Behavior

Play behavior is a system that offers many opportunities for trial-and-error learning; hence, it will be discussed in this context. As with a number of other behavior categories, play behavior is debated with respect to its limitation and interpretation. It is not known whether play constitutes one unitary phenomenon, and a generally acceptable definition for play cannot be given. Instead, we will describe some characteristics and regularities of this category.

7.4.4.1. Criteria, Form, Classification, and Occurrence

PLAY BEHAVIOR can be considered "behavior that is not serious." In play, new behavior patterns are invented, or behavior patterns or segments occur that in other contexts have serious purposes, e.g., escape, attack, or prey catching. The behavior patterns

- ☐ appear "without purpose" since they do not fulfill the biological function that developed during their phylogenetic history.
- ☐ can be variously combined, i.e., not only is there no mutual inhibition but the strict temporal correlation of individual behavior patterns in the framework of behavior sequences is absent as well (see Chapter 4).
- ☐ do not seem to be subject to habituation as far as their release is concerned (see Section 7.4.1), since many behavior patterns occur during play for considerable periods of time without any sign of fatigue.
- ☐ appear frequently "exaggerated" in comparison to their "serious" execution, i.e., they are performed with greater amplitude, vigor, and speed, or they are repeated more frequently.

A classification of play behavior patterns in animals can be made according to two criteria, depending either on the object or on the kind and number of partners that are involved. In the first category are movement, object, and social play, depending on whether an animal plays with an inanimate object or with a conspecific. In the second category, differentiation is made between play involving siblings, parents, and their young and play involving several juvenile or adult conspecifics.

Play behavior occurs only in the most highly evolved vertebrates and birds. It is widespread in mammals, but rarer in birds, where it is expressed strongly only in the Corvidae. This difference between mammals and birds may be due to the faster juvenile development in birds, which allows little "time" for play behavior. Some instances of play have been described in fish and reptiles, but it is clear to what extent it is like play in higher vertebrates.

As a rule, play is limited to young animals or occurs most frequently in them. In animal groups that are very responsive to their environment, play behavior—and especially curiosity—is present even in adults. This is true especially for predators, rodents, primates, and whales, and porpoises show extensive play behavior as adults.

There are indications that the soft, highly variable "subsong" of some juvenile songbirds shows parallels in its form and function to "true" play behavior. Hence, it seems justified to consider it play and call it "play song."

7.4.4.2. Function

Play behavior involves a high expenditure of energy and increased vulnerability (being more conspicuous, possible weakening and injury). The fact that play behavior is frequent and widespread in some species, and that many highly developed behavior patterns and specialized forms of communication occur, leads us to suspect that it has a biological significance that compensates for the disadvantages listed above. Hence, we suspect a selection pressure that may have led to the evolution of play behavior.

In principle, we can distinguish two functions that cannot always be clearly separated: first, the immediate benefit for the growing organism, and second, the long-term significance for the adult. For the first, only cautious conjectures are possible. It is possible that that play behavior benefits the growth of muscles, and that sensory impressions obtained during play favorably influence the maturational processes of the sense organs and the central nervous system. In other words, there seems to be a connection between play and the speed of growth and development.

The long-term significance of play may be

- motor exercise, i.e., practice of muscle functions.
- influence on the process of socialization, i.e., the learning of social roles, including the ability to recognize individual social partners, the development and improvement of social communication, the control of the individual's aggression, and the development of social bonds.
- cognitive development, i.e., the practice and improvement of overall perceptual abilities (see footnote 1 in Chapter 12.4.2).

Of course, these three possible functional systems cannot be sharply separated from one another. Most likely all play behavior does not have the same significance for all three systems, although in all instances play behavior can aid in the acquisition of experiences in the widest sense. A learning capacity has to be presumed, and investigations on various species have shown that the readiness to learn is especially great in young animals.

7.4.4.3. Motivation and Communication

There is still a question about the motivation for play. Play behavior occurs only in a so-called relaxed field (state), i.e., at a time when no other behavior tendencies are activated. This, the ease with which behavior elements from various functional systems can be combined, and other characteristics of play make it seem unlikely that the motivation of a behavior as part of and outside of play are the same in a specific instance. There are some indications that there may be a specific play appetence (see Section 2.8). Hence, it seems possible that play has its own motivation, e.g., a play or curiosity drive. However, proof for this hypothesis is still required.

Some species that have a highly developed play behavior have evolved their own play signals. They indicate their readiness to play, and show signals that seem designed to avoid "misunderstandings," especially during attack games, which indicate to the playmate that certain aggressive behavior patterns are not meant "seriously." A well-known example for play signals is the "play face" found in primates and some predators, where it can be interpreted as a ritualized intention to bite (see Section 10.6).

7.4.5. Imitation, Social Facilitation

Many animals are capable of imitating behavior, usually of a conspecific. Hence, they gain experiences in this indirect manner. This is called IMITATION or OBSERVATIONAL LEARNING. Imitation is the incorporation of observed movements or perceived vocalizations into the motoric or the vocal repertoire of the individual.

Acoustic imitation occurs especially in songbirds and parrots. If one species imitates the vocalization of another, as is the case with, e.g., starling, icterine warbler, marsh warbler, and mockingbird, it is called MOCKING. In some species, e.g., in the Australian lyrebird, this tendency to imitate other species is so strong that the other species' vocalizations predominate in the vocal repertoire of the imitator. The biological significance of mocking is still not understood, despite various attempts to explain it.

Motor pattern imitation involves higher demands, since the animal's own movement may be substantially different from that of the model and is only known with certainty in some mammals. It is extensive only in primates, where imitation has also been observed in the wild. Anthropoid apes, especially chimpanzees, are capable of imitating fairly complicated behavior sequences spontaneously, i.e., without prior trials.

One phenomenon that is frequently confused with imitation is mood induction, although it can be clearly distinguished. MOOD INDUCTION, or "contagion," is a widespread tendency in social animals to do the same thing. A satiated hen is stimulated to peck at seeds when observing eating conspecifics. An English sparrow is stimulated to dust-bathe by the example of dust-bathing conspecifics. In flocking birds, one group may simply "sweep along" others not yet in a flying mood. Many preening or grooming movements and vocalizations (e.g., barking in dogs) are contagious in this respect.

Superficially, mood induction, an important means of synchronizing behavior in a group under natural conditions, may appear to be imitation. Actually, an animal is only placed into a "mood" to perform a behavior that it already "knows." Hence, no acquisition of new information, in the sense of a learning process, is involved. In true imitation the animal acquires a new ability, i.e., it takes on movement patterns or vocalizations that it had not mastered previously.

7.4.6. Insight Learning

A very impressive learning performance, found only in primates—and also in the form of detour behavior in some other species—is learning by

INSIGHT. It is characterized by the fact that an animal comprehends a new situation spontaneously, considers the necessary spatial and temporal behavior sequences in advance, as it were, and then performs the sequence correctly the first time.

Methodologically, however, the demonstration of such abilities presents great difficulties. The prior history of the experimental animal must be known so as to exclude any opportunity to acquire any of the necessary experiences in similar situations. Furthermore, the experiment must be designed so as to prevent any opportunity to practice, and to enable the animal to view the entire situation prior to the test, so that it can perform the required solution first "in its mind." In spite of this, a chance solution cannot be excluded in the individual case because one cannot see the "insight" taking place in the animal. If, however, the animal repeatedly solves various problems without prior practice, then a chance solution is not the explanation. In addition, there are a number of criteria that may indicate whether the animal did, in fact, comprehend the solution prior to doing so.

Extensive pioneering experiments on insight learning were carried out in the 1920s by W. Köhler on chimpanzees. He presented the animals with food that was placed out of their reach, along with a number of objects (rods or tubes that could be joined, boxes that could be stacked, etc.) that enabled the animals to reach the food only if they arranged them in certain combinations. It was found that the animals were able to use various new forms of behavior

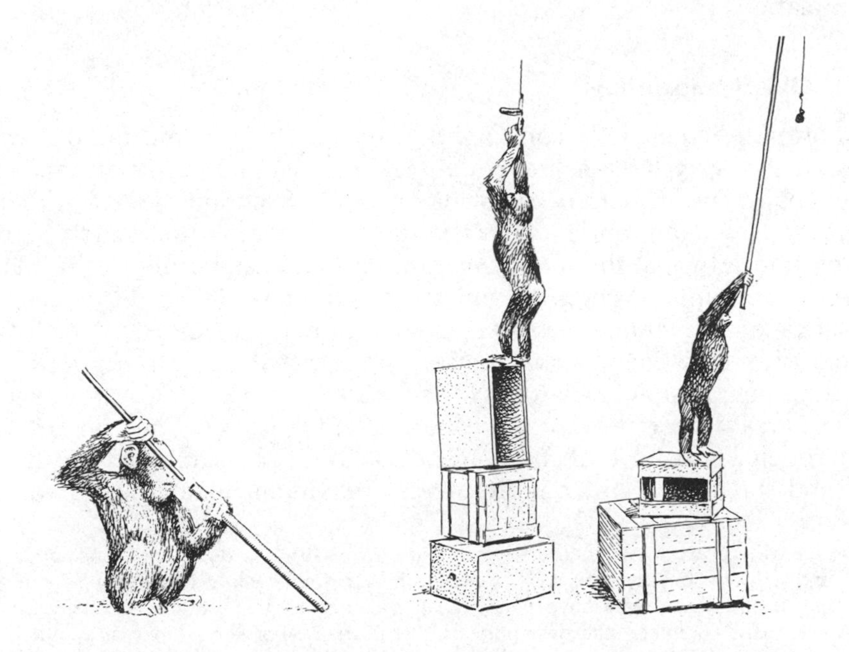

Fig. 43. New combinations of behavior in chimpanzees. To reach otherwise unobtainable food, they join sticks and stack boxes (after Köhler 1921).

in the solution of the tasks. This ability was confirmed in increasingly complex experiments. Some of the solutions to problems approached human capabilities.

The simplest form of insightful behavior can be tested in the so-called detour experiment. Here, insight into the spatial situation can be assumed whenever the experimental animal performs a detour toward a goal when the direct approach has been blocked and where the animal must first move away from the goal, to approach it later only after the obstacle has been circumvented. The initial running in the wrong direction makes sense only as part of the total detour, and can only be understood from insight into the total situation. Spontaneous detour behavior is known especially in mammals (dog, badger, rat), in a few instances also in lower vertebrates (e.g., the dwarf chameleon), and even in insects (e.g., the digger wasp *Ammophila*).

7.4.7. Imprinting

In spite of the large number of investigations and discussions, the process of IMPRINTING is still controversial as far as the interpretations are concerned. Some authors see it as a special learning process that is more or less different in principle from other learning processes, while other do not accord it separate status and consider it a form of conditioning in a wider sense. Finally, there is some indication that we are not dealing with a singular phenomenon and that the unique features of imprinting are not to be found in the acquisition of information.

7.4.7.1. Object Imprinting

Imprinting is generally considered to be a very early and rapid learning process with a very long-lasting result. The classical examples of imprinting refer to "objectless" inborn behavior patterns (see Section 6.1.3) for which the knowledge of the appropriate object is thus acquired. A distinction is made between imprinting of the following response and sexual imprinting. In the first instance, which occurs in the young of precocial bird species,[3] the knowledge of the mother or parents determines tbe knowledge of those characteristics by which later on the sexual partner is recognized.

Young precocial animals follow the mother or parents closely, a reaction that is of great importance for the cohesion of the family. They possess only a very rough knowledge of the "object" of this following reaction: It must move and have a certain minimum and maximum size. During the first

[3] Species are called PRECOCIAL when their young are born with fully developed sense organs and limbs, and are able to leave the nest or place of birth immediately or shortly after birth or hatching. ALTRICIAL animals have young that are born at a substantially earlier stage of development, and complete their development later in the nest or den. They require much more intensive care by their parents. Originally this distinction was made only for birds. Precocial animals include ostriches, chickens, ducks, hares, and hoofed animals. Altricial animals include raptors, pigeons, songbirds, and most predatory mammals and rodents.

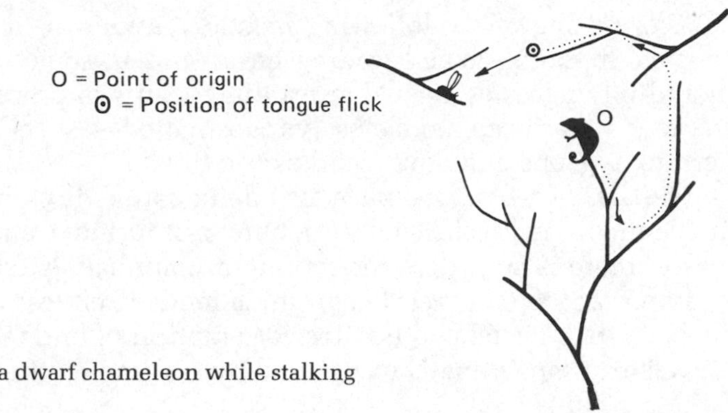

Fig. 44. Detour behavior in a dwarf chameleon while stalking prey (after von Frisch 1962).

following reaction they are imprinted on additional cues, and from then on they prefer to follow an object that possesses a complete array of these characteristics. Thus, if one places duck eggs under a hen for incubation, the young duckling will, after hatching, imprint on the hen's characteristics and follow her until the waning of the following response. Young birds hatched in an incubator can be imprinted on any number of species—humans, moving models such as balls, boxes, etc.—with permanent attachment to these objects being the result.

The knowledge of the characteristics of the sexual partner is in many species, including birds, very incomplete or inborn only in rough outline. If young animals of such species are raised by foster parents of another species, then they prefer to mate as adults with members of the foster species. In many instances this malprinting may last a lifetime. This is called SEXUAL IMPRINTING. When animals have been hand-raised by humans, one speaks of HUMAN-IMPRINTED ANIMALS.

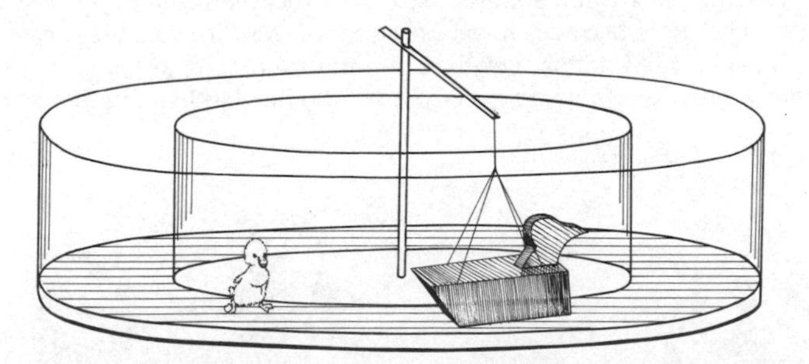

Fig. 45. The apparatus originally designed by E. H. Hess for the study of imprinting of the following response in young precocial birds. A model, suspended from a rotating arm, moves across a circular runway. A loudspeaker can be fitted inside the model. The ducklings, which were kept in the dark from hatching until the beginning of the experiment, are exposed at various ages and for varying lengths of time to the model. Thus, it is possible to determine the beginning and end of the sensitive phase of imprinting, as well as the amount of necessary experience with the model. The importance of the mother's call, the influence of obstacles, aversive stimuli and drugs, and many other questions can be tested in this apparatus.

Imprinting of the following reaction is known in many duck and goose species, in chickens and wading birds, and in some other precocial species, including ungulates. Sexual imprinting occurs in precocial as well as altricial animals. It has been more closely investigated—especially in ducks, chickens, parrots, pigeons and doves, and exotic finches.

Initial investigations have also demonstrated sexual imprinting in fishes. In mammals many instances of human-imprinted animals are known. The reason there is so little information on mammals is that olfaction plays such an important role in their lives, which lends itself less readily to experimental analysis than visual and auditory orientation of birds.

Object imprinting is recognized by two main criteria:

- ☐ It takes place only during an early, sensitive phase (see Section 7.2), which is especially short, and with respect to the following reaction may last only a few days or even hours. In sexual imprinting this sensitive phase is limited to a few days or weeks.
- ☐ It results in a very lasting bond. This means that the object preference that was acquired during the sensitive phase can be changed later in life only with great difficulty or not at all.[4] This long-lasting effect is referred to as irreversibility.

In addition to these criteria, there are a number of others that are used to characterize imprinting phenomena. Hence, it is typical for sexual imprinting to take place early in life long before sexual maturity is attained, i.e., *before* the first expression of the behavior. This criterion applies only to sexual imprinting. In imprinting of the following reaction, the object preference is accomplished directly only during the actual following itself. Other criteria

[4] It must be emphasized that only a *preference* is established. If the animal is later tested with only one object to which it was not exposed earlier in life, then it may court or follow it as the threshold for sexual behavior becomes lower (see Section 2.2). Irreversibility can, therefore, only be tested in a free-choice situation between the imprinting object and other objects. This precaution has not been taken in many experiments and has led to a number of misunderstandings.

Fig. 46. A zebra finch that was raised by another species (society or Bengalese finch) courts the model of a society finch in the presence of a model of its own species.

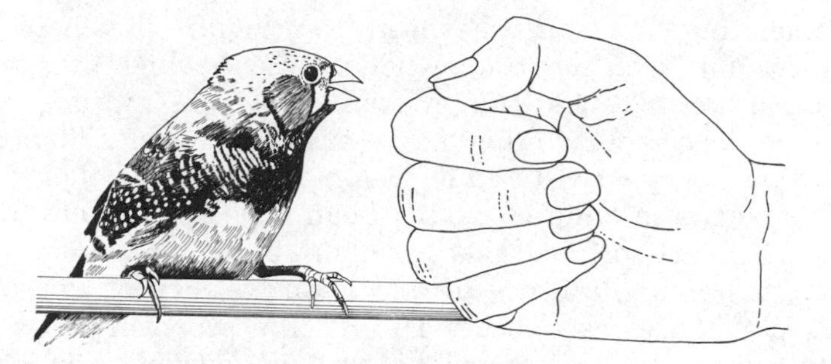

Fig. 47. A hand-raised, and thus human-imprinted, zebra finch courts the hand of the caretaker.

apply only to the following reaction, but seem to have only limited validity. The basic criteria for imprinting remaining are only the sensitive phase and irreversibility.

7.4.7.2. Imprintinglike Processes

In recent times, it has been discovered that there are additional developmental processes that take place in an early and limited period, and that lead to a more or less stable result, i.e., to which the above criteria for imprinting also apply. Depending on the degree of correspondence, one speaks of imprinting or of "imprintinglike" processes, although the concepts are not specifically defined.

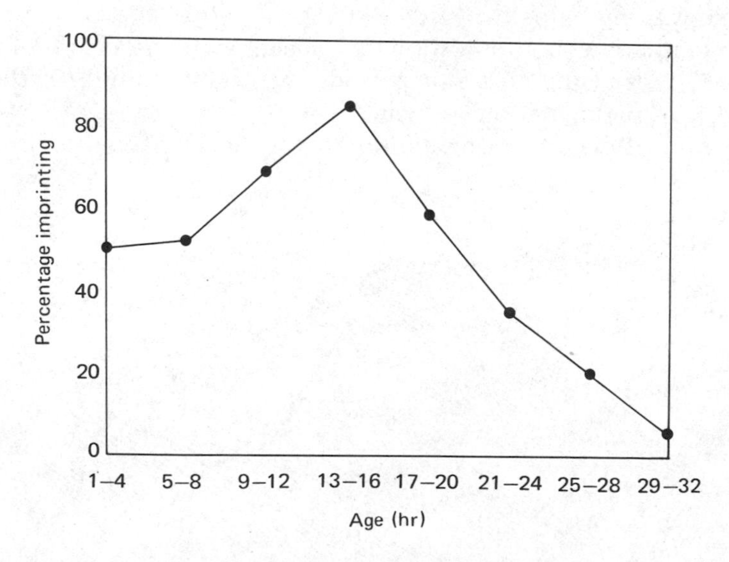

Fig. 48. Investigations of mallard ducks show a temporal limitation of the sensitive phase for the following reaction. The points on the graph indicate the percentage of chicks in the appropriate age group that imprinted on a model during a 10-minute training session, i.e., those which in a free-choice test followed a familiar over an unfamiliar model (after Hess 1959).

Many migrating birds and fishes learn to recognize the characteristics of their birthplace during an early ontogenetic phase to which they return again (LOCALITY IMPRINTING). Pacific salmon, for example, are imprinted to the odor of the stream or lake in which they are born, which enables them to find the same place where they spawn two to five years later upon their return from the sea. This occurs in spite of having been exposed to many other odors since then. It was demonstrated that no inborn preferences were involved by transplanting the young. With respect to FOOD IMPRINTING, like, rapid, and long-lasting preferences can occur. This is true especially in species that parasitize certain plants, e.g., insects that are monophagous and whose larvae live on only one plant species.

A well-investigated case is the so-called HOST IMPRINTING of the African wydah birds. The species of this subfamily of the weaver birds are all brood parasites that lay their eggs among those of grass finches, who then raise their young. One wydah species parasitizes only one specific grass finch species. In some instances, specific subspecies of wydah birds use only specific grass finch subspecies as hosts. This specialization is necessary since, unlike the European cuckoo, the young wydah birds do not eject their host's nest mates from the nest but grow up with them together. In adaptation to this, they developed during the course of evolution an almost complete correspondence in appearance and behavior with the young of their respective hosts, which is quite in contrast to their appearance as adults and their behavior. This host-specificity is apparently maintained by an imprinting process: the wydah birds imprint on the appearance and vocalizations of the host species. When they are adults, they find, based on their early experience, their host species when these are engaged in reproductive activities. This results in two goals: One, host imprinting helps the birds to locate a pair of the host species and results in a temporal synchronization. It appears as if the nest-building and courtship activities of the host pair provide important stimuli for the female wydahs in triggering ovulation in them.

At the same time, this orientation to the host species facilitates the

Fig. 49. A male wydah bird (top) attracts a female (bottom) to a nest-building male of the host (violet-eared waxbill) with the nest call of the host species (after Nicolai 1964).

meeting of sexual partners with the same background, i.e., the same host-imprinting experience. An important factor in this is the fact that the wydah male has acquired the entire song repertoire of the host species and possesses it along with its own. Hence, his song contains elements that a young wydah female heard from her grass finch stepfather.

Beyond these object preferences, other permanent influences on early experience have been discovered in the development of other behavior characteristics. This is true for motivation: In some species (mice and rhesus monkeys), the level of aggression in adulthood seems to depend on the extent of the animal's individual aggressive experiences early in its life (see Section 8.1.9.3). Similar results exist for the need for social contact in various social species. In dogs the capacity for socialization later in life is determined during a sensitive phase (SOCIALIZATION PERIOD) between the third and fourteenth week of life (with a peak during the seventh week). Finally, many social maladaptations that are known under the heading of deprivation syndrome can be traced back to situations in which the young of social species are raised in isolation, which has negative effects when an experience occurred during a very specific developmental period (see Section 12.3.1).

7.4.7.3. Motoric Imprinting

In some instances, the performance of movement patterns can be permanently determined through early experience. This is especially true for the song-learning of some birds: In the North American white-crowned sparrow and in the Australian zebra finch, the male learns his song quite early, i.e., before he begins to sing himself. The young male acquires by listening, usually to his father, a kind of template against which it matches its own song development later in life. The parallel with imprinting lies, on the one hand, in the relatively short, sensitive phase for this learning process, which in the white-crowned sparrow is between the thirtieth and one hundredth day of life. The other similarity is the permanence of the result: after the sensitive phase, the animals are unable to learn, and the song is finally determined. This can be demonstrated with exchange experiments: If one has another finch species rear young zebra finches, they learn their song. Later on they show no additional changes in the song of the foster species, even if they live for years thereafter only with their own species.

A review of the relevant literature shows that early ontogenetic experiences, whose biological significance was already mentioned (see Section 7.2), play an important role in many functional systems. It also shows the variability of learning processes that have been referred to as imprinting. Nevertheless, the two general criteria of imprinting apply to them. If one examines these criteria more closely, it becomes clear that they have less to do with the learning process itself than with the storage of the information in memory, i.e., the fact that a certain amount of experience during a specific developmental period leaves a more permanent impression than the same experience

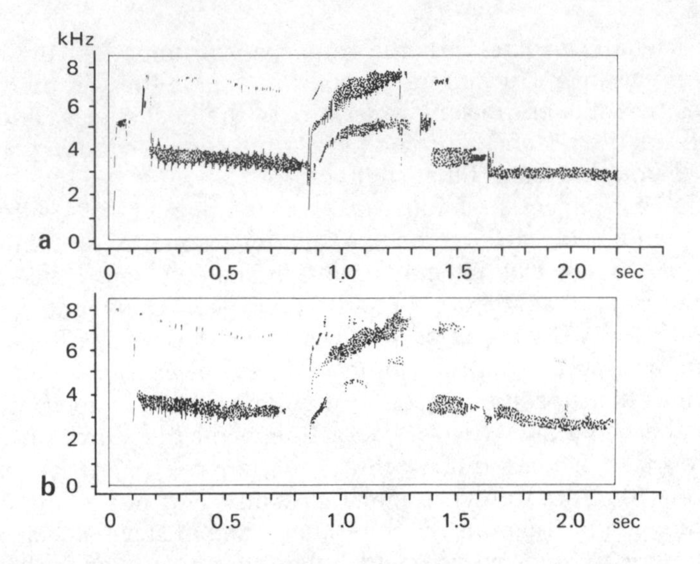

Fig. 50. Imitation of the host species' song by a wydah bird. Top: Sequence of a host species' song, a Melba finch. Bottom: The same sequence from the song imitation of the paradise wydah bird (after Nicolai 1964).

at a later time in life. Perhaps what we mean by imprinting is more of a memory than a learning phenomenon.

It can be postulated that imprinting is correlated with certain developmental processes in the brain. We know almost nothing about its morphological and physiological basis.

7.5. Behavioral Traditions

Two behavioral systems will be discussed here because they are closely correlated with various learning processes, although they do not depend in all instances on learning processes: They are tradition with respect to behavior and to tool using.

By tradition we mean the transmission of acquired, i.e., learned, information. This can be directed toward two goals: in social animals the information can be passed on within the group, and in species from one generation to the next.

Four kinds of tradition can be distinguished in two categories: direct and indirect tradition, and object-dependent and object-independent tradition. In OBJECT-DEPENDENT TRADITION, "sender" and "receiver," i.e., experienced and inexperienced partner, must meet at the same time or shortly thereafter with the "object", e.g., food, about which knowledge must be acquired or whose preparation must be learned. A famous example of such a tradition is the "potato washing" in a population of the Japanese macaques on the island of Koshima. This technique involves the dunking into the water of a potato and wiping it with the other hand to free it from dirt and sand. This was "invented" in 1953 by Imo, a young female, and gradually spread over the entire island population. Many other practices vary from one group to the next in this species, but they are uniform within a group, due no doubt to tradition. It is interesting to note that existing habits are apparently passed on from parents to young or from older to younger group members, while *new* habits spread in the opposite direction, i.e., they are usually first "invented" by younger animals.

The expansion of behavioral characteristics by tradition is also known in other vertebrates. It is known that Norway rats learn from group members to prefer or avoid certain kinds of food. In England, tit mice learned toward the 1930s to open the tinfoil covers of milk bottles, which enabled them to reach the cream layer on top of the milk. This ability appeared independently in several locations and spread rapidly from the various points of origin. Besides these feeding traditions, place preferences can be passed on from one generation to the next. Thus, bighorn sheep in North America migrate between their scattered grazing areas on paths that the species already knew at the end of the last Ice Age, even though many of these paths lead through recently forested areas that are quite unsuitable habitat for the animals. The preference for certain localities in many migrating birds and fishes offers other examples for such traditions.

Finally, the knowledge of species-typical vocalizations can be passed on from one generation to the next by tradition when the young birds learn the song from their father. In many songbird species such SONG TRADITIONS are known. An impressive example of the stability of such traditions is provided by some European songbird species that were settled about 100 years ago in New Zealand. In recent studies, using tape recordings and sound spectographic analysis, it was shown that the songs of the New Zealand population still agree with those of the European population from which they originated. It is known from at least two of the investigated species (European blackbird and chaffinch) that they must learn large parts of their song. Thus, the learned portion of the song has been passed unchanged through the generations from the original population despite a separation of 100 years. Tradition also plays an important role in the development and transmission of dialects in the songs of birds (see Section 8.2.2.5).

In contrast to object-dependent transmission of information by tradition are those situations in which information about a specific object, e.g., the location and nature of a food source, is transmitted even in its *absence* (OBJECT-INDEPENDENT TRADITION).

However, this process requires certain symbols for identification. This possibility was actualized in the "dance language" of honeybees.[5] It enables forager bees that have returned from visits to certain flowers to inform their

[5] The term LANGUAGE is used somewhat differently in ethology from everyday usage. Language here does not imply words or even vocalizations, but includes *any kind* of communication in which symbols are used.

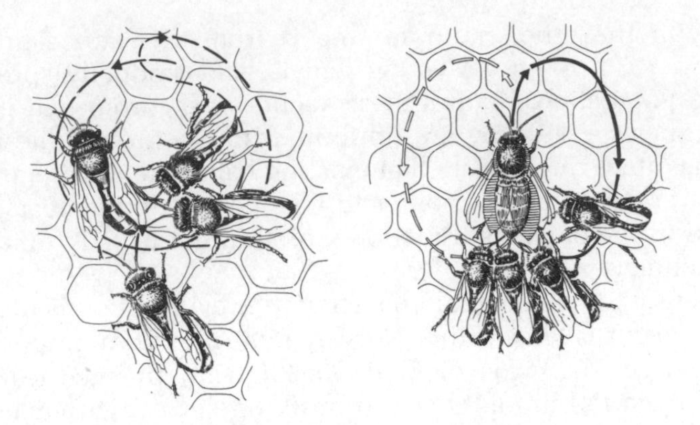

Fig. 51 (left). The round dance of the honeybee. The dancer is followed by three workers (after von Frisch 1965).

Fig. 52 (right). The waggle dance of the honeybee. The vertical portion of the path is where the waggle dance is performed. Here, too, several workers follow the dancer (after von Frisch 1965).

hive mates about the location, distance, and richness of a food source by means of distinct movement patterns: The so-called round dance is seen when the food source is near the hive. Its function is simply to excite other foragers by rapid and conspicuous running in a circle on the comb, and to incite them to follow her. The specific information about the food source that they can use consists of the odors that adhere to the dancer and that the hive bees detect with their antennae. The waggle dance is able to convey information about more distant food locations. The bees are able to perform remarkable feats of transposition, where the location of the food source with reference to the sun—which cannot be projected directly on the vertical combs—is indicated with the aid of gravity: If the dancer orients the direction of the waggle portion of the dance upward, then the food is located in the direction toward the sun. If the waggle dance axis deviates from the vertical to the left or right at a certain angle, then the flight path to the food also deviates by that same angle from the straight line to the sun. The speed with which the waggle portion of the dance is traversed indicates the distance to the food: The greater its distance from the hive, the slower and the more exaggerated the movements of the waggle dance. The transition from the round to the waggle dance lies between 50 and 100 meters from the hive.

A similar, non-object-related transmission of information can be experimentally demonstrated in chimpanzees: These animals are unable to articulate a large number of vocalizations. Hence, they have a limited ability to communicate verbally. However, they can learn the sign language of deaf people and are now able to communicate better with their caretakers. What the sign language used by the chimpanzees has in common with the dance language of bees is the use of signs that permit the transmission of the information in the absence of the objects in question.

The second classification of traditions refers to the relationship of the partners involved: In the DIRECT form the inexperienced partner learns directly from the parents or other group members. In the INDIRECT form, as is the case in some parasitic insects, information transmission consists merely of the female laying her eggs on a specific host plant or animal; as a result, the larva's preference for the host species becomes established so that, as an adult, the insect again selects the same host for egg laying. Hence, the preference is passed on from one generation to the next without the two generations ever meeting. The only prerequisite for this kind of tradition is a certain amount of brood care (see Section 8.3.1).

Indirect traditions are always dependent on the presence of objects, whereas in direct transmission of information in reference to an object, both possibilities can exist.

All traditions discussed so far involve interaction between members of one species. However, there are also instances of so-called INTERSPECIFIC TRADITION. The best-known example is that of the male wydah birds, which learn songs of their host species in addition to the most likely innate component of their species' song (see Section 7.4.7.2). In this example, information is passed on new in each generation from one species to the other.

Traditions may consist of a rigid code system, whose components are inborn, as is the case with the dance language of bees, or they can involve imprinting processes, as in many habitat or food traditions. Finally, they can be based on learning by imitation, as in the potato-washing technique of the Japanese macaque.

7.6. Tool Using

Some species have the ability to use objects as "tools" when grooming, in obtaining and handling food, and in other situations. Elephants take up sticks with their trunks to scratch themselves on their backs; Egyptian vultures break open ostrich eggs by throwing rocks at them; chimpanzees use sticks to hit

Fig. 53. A woodpecker finch with a cactus spine in its bill. The tip of the spine is oriented into the burrow of an insect larva (after Eibl-Eibesfeldt and Sielmann 1962).

stuffed leopards, or they stack boxes to reach bananas suspended high above them (see Section 7.4.6). A much-cited example of tool use is provided by the Galapagos woodpecker finch, which uses a stick or cactus spine in its beak to probe for hidden insects. Weaver ants take their larvae, which have spinning glands between their mandibles, and use them like spindles between two adjacent leaves on both sides of their thorax. The result is that the leaves are woven together by a dense webbing. Thus, the animals use "living objects" as tools. As adults they lose this ability to spin.

Some animals can either manufacture or improve their tools. Chimpanzees break off branches, strip off the bark, break off small branches, and use them to probe for termites in their hills. The woodpecker finch can also break the cactus spine to the right length.

Tool using can have an inborn basis, e.g., with the weaver ants. However, in most instances the use of tools seems to be learned. This can occur by trial and error, e.g., with the Egyptian vulture, but it can also be the result of insight and not merely experimentation, and appears to be "planned," as observed in chimpanzees. Sometimes a specific use of tools can be passed on by tradition.

8

Social Behavior

Only very few species are socially completely indifferent and never exhibit any relationship between conspecifics. They include primarily some lower marine organisms, which are sessile on the substratum and which disperse their sexual products (eggs and sperm) into the water. However, even here a temporal synchronization by chemical substances can occur between neighbors, and this can be considered a form of social contact. In the overwhelming majority of animals there are, however, frequent encounters with conspecifics, especially during the reproductive phase (courtship and care of young), but also during other times, either permanently or in certain periods in the animals' lives or the time of year.

Although much of what was discussed in previous chapters also applies to social behavior, there remain many regularities that are exclusively a part of this functional system. We will present them here in a separate chapter so that their interrelationships may be better appreciated.

The concept SOCIAL is applied in ethology without attaching to it any special value judgment. Hence, it should not be taken with the positive connotation of the word when applied to human society. The concept merely states that the behavior patterns associated with it serve the communication between species members. In this sense, social behavior includes all behavior between conspecifics.[1] For this reason, intraspecific aggression, of biological importance in several respects, is also considered a part of social behavior. It will be discussed first.

8.1. Fighting Behavior

8.1.1. Definition

The functional system of fighting behavior comprises two parts: attack and escape. Attack behavior in its widest sense is often referred to as AGGRESSIVE BEHAVIOR, while the underlying motivation is called READINESS

[1] The use of the word SOCIAL is not the same throughout ethology. As a rule, all behavior patterns shown between members of one species can be considered social. However, some authors use the concept in a narrower sense and limit it to behavior occurring within a group of animals. Social or gregarious species are all those who live in permanent pairs (see Section 8.2.3), in families, in groups, or—in insects—in colonies, and where social behavior constitutes a large part of their total behavior. Solitary animals are those who largely live alone. Examples among mammals are the sloth, many predatory cats, and the European hamster.

TO ATTACK, AGGRESSIVITY, or AGGRESSIVE DRIVE (see Section 8.1.9.5). The term AGONISTIC BEHAVIOR used mainly in the English literature, is not identical with these terms. It is more of a superconcept and refers to *both* components, including all of those behavior patterns that are made of a combination of attack and escape, e.g., threat behavior.

Aggressive behavior can occur between conspecifics as well as between members of different species. Interspecific aggression includes defensive reactions and competitive behavior, e.g., at a watering hole, vultures feeding, or at a nesting site—as in some hole-nesting species of birds. Some authors also include prey-catching behavior of predators among interspecific aggression, while others consider this simply as eating behavior. Intraspecific aggression, however, is exclusively the result of competition, where the objects of competition include not only part of the nonliving and living environment (home range, hiding places, mating arena, egg-laying location, food, symbiotic partner) but also conspecifics, especially the sexual partner.

On several occasions it has been discussed whether or not there exists a causal relationship between the two forms of aggression, i.e., whether intraspecific aggression was derived from prey-catching behavior or from defensive reactions against predators, and hence from interspecific competition. The hunting and killing of prey animals has frequently been considered to be the phylogenetic origin of intraspecific aggression. Many facts argue against such a direct link. Many plant-eaters, which are often contrasted with predators as being "peaceful" animals, are characterized by a high degree of intraspecific aggression. Furthermore, the behavior patterns involved in intra- and interspecific aggression can be quite different (see Section 8.1.7). Finally, neuro- and hormonal physiological investigations show that both can be elicited at different stimulus sites by brain stimulation (see Section 5.1.2.2), and that they can be differentially affected by hormones. There is very little evidence to the contrary. One positive correlation has been found in two species of baboons in which the intensity with which a male fights for possession of a female is greater in the species in which the male also defends his group more vigorously against predators. All currently available evidence does not allow us to conclude that generally a predatory life-style was a prerequisite for the development of intraspecific aggression, which has been argued with respect to humans on several occasions. It appears more likely, because of the various differences that can be observed, that both forms of aggression developed separately in response to different selection pressures during the course of evolution.

From an ethological perspective, intraspecific aggression has aroused more interest because it has been the source of very highly developed behavior patterns. The following sections deal with the various forms and regularities of intraspecific fighting behavior.

8.1.2. Biological Significance

The biological significance of intraspecific aggression is manifold. First, it insures that members of a species distribute themselves singly, in pairs, or in groups more or less evenly throughout the available habitat. Hence, they optimally utilize the available food base. Furthermore, it can insure that surplus individuals will migrate when the population density in a particular habitat has been surpassed *before* an actual food shortage weakens the population[2] as a whole. At the same time, this facilitates the expansion into still unsettled habitat. Hence, intraspecific aggression disperses competing conspecifics and guarantees the spatial prerequisites for reproduction, inhibits the spread of disease by this "thinning out," and offers in those species with protective coloration a better protection from predators. It also aids in sexual selection by selecting the strongest and healthiest individuals for reproduction. In social species it can assure leadership by the most experienced individuals by the development of a rank order. Of course, not all functions are of equal importance in all species, and in individual instances there may be additional functions of intraspecific aggression.

These various biological functions entail a number of disadvantages for the individual and for the species: Fighting animals can become weakened or injured, they are more in danger because they are more conspicuous and less alert, and they lose time inadvertenly for other activities. Theoretically another danger may be that growing youngsters that are physically inferior and inexperienced by comparison may be adversely affected, and hence the species may lose its reproductive reserve. For this reason the extent and form of intraspecific aggression in a species developed out of conflicting selection pressures that derived from the needs for camouflage and self-preservation of the individual, on the one hand, and the density-regulating and selective requirements for the species on the other.

This compromise is evidenced from one species to the next in different levels of aggressive motivation (see Section 8.1.9.1), and also in the fact that in many species behavior patterns have evolved that insure all the advantages of aggression but avoid most of its disadvantages. This includes threat and display behavior, submission and appeasement postures, territoriality, individual distance, and finally, the "invention" of largely harmless forms of combat.

8.1.3. Threat and Intimidation Displays

Behavior patterns that can fulfill the same drive function as can actual fights, in which the participants are less endangered—and hence they largely prevent the occurrence of injuries among the rivals—are included in the category of THREAT BEHAVIOR. Their biological function is in the intimidation

[2] By a POPULATION is meant the total number of animals of a species in a certain area. POPULATION DENSITY is the average number of individuals per unit of area. It is regulated by a number of environmental factors, such as availability of food, predation, and competition from species with similar requirements, and is unique to each species within certain limits. Due to the availability to prey, this density is usually less in predators than in plant-eaters, and it is less in larger than in smaller animals. In many species, more or less regular fluctuations of their population density occur. Frequently they correlate with the seasons, but they can also span longer time periods. An example for a population cycle over several years is the northern lemming.

Fig. 54. Anolis lizard in normal body position and in a threat posture. During the threat the silhouette of the animal is greatly enlarged by the inflation of the throat sack and the erection of skin folds (after Kästle 1963).

of the rival, who is induced to turn away before an actual fight commences.

Two forms of threat behavior are especially frequent: either the threatening animal enlarges its outline by erecting the fur or feathers, spreading skin folds or gill covers, and inflating a throat sac, or it presents its "weapons" (teeth, beak, horns, or antlers) conspicuously. This can be combined with intention movements (see Section 10.6), which are a part of the actual fighting behavior. Frequently these movements are in conjunction with special color patterns.

Motivational analysis has shown that threat behavior can be based on a mixture of motivations of attack and escape tendencies (see Section 2.8).[3] This can be recognized in many instances by the "ambivalent" form (see Section 4.2) of the behavior patterns in question, e.g., their incomplete attack or escape components. This causes the extreme thickness of a threatening black-headed gull's neck by the simultaneous contraction of all muscles, i.e., those that would enable the bird to thrust its beak toward the opponent and those that pull it back. Some species, e.g., cats or gulls, possess several threat postures, which in a specific instance are composed of various components of the attack and escape behavior.

Some movements and body postures of threat behavior have a dual function: They have a repelling or intimidating effect on rivals of the same sex, but they attract members of the opposite sex. Behavior patterns that frequently include courtship patterns are called DISPLAY BEHAVIOR. They occur mostly among males. The decision as to whether display or pure threat behavior is shown can, of course, only be made in cases where the function of a behavior is well known. In the literature, the two concepts are not always clearly separated, and the word *display* is often used synonymously with threat behavior.

The significance of a coupling of rival intimidation and attraction of a sexual partner is obvious: Threat behavior occurs frequently during the reproductive period, especially during the establishment of territories (see Section 8.1.4). It is a possible signal for a reproductively motivated, territory-holding male, and it seems appropriate that an unmated female should respond. Display behavior as such plays an important role during courtship, and especially during the pair-formation stage of many animals.

[3] This statement is based on analyses with birds and fishes. How applicable it is to mammals has not been determined as yet.

8.1.4. Submission and Appeasement Gestures

Additional behavior patterns that, like threat behavior, can contribute to the avoidance of harmful effects of intraspecific aggression are submissive and appeasement gestures. Both inhibit aggression, but they differ in form and function. A SUBMISSIVE GESTURE is usually the direct opposite of the species-specific threat behavior: Body size is decreased by depression or folding of movable appendages; dangerous weapons and other releasers of aggression are concealed or are "demonstratively" turned away, while especially vulnerable parts of the body (throat, back of the head, neck) are sometimes actually turned toward the opponent. Animals that can rapidly change their body coloration—e.g., squid, cichlids, some lizards—can remove the colors that are releasers of aggressive behavior from view by the opponent by appropriate color changes, even without the use of special movements.

Submissive gestures are exhibited in intraspecific fights by the losing animal, and the ability to inhibit further aggression is so strong that the fight is at the least temporarily discontinued. Appeasement behavior is especially developed in well-armed animals, and they occur in situations in which escape is not possible, e.g., in members of a closed group where leaving the latter would entail risks that would be biologically unacceptable.

The effect of an APPEASEMENT GESTURE—in contrast to a submissive posture—is not to "turn off" previously exhibited aggressive signals, but the activation of other behavior tendencies that are incompatible with aggression. Usually the behavior is derived from the parent–young relationship and from sexual behavior. It is used as a "greeting" in species that live in pairs or groups, and makes possible closer approach and living together.

A sharp distinction between submissive and appeasement gestures is not always possible, since elements of both can occur in combination. Hence, some authors do not make this distinction and use both concepts synonymously.

8.1.5. Territoriality

The damaging effects of aggression can be avoided still more effectively when even the comparatively "harmless" threat behavior is temporarily

Fig. 55. Three different threat postures of the herring gull. Position (a) indicates a low tendency to flee, (c), great readiness to flee (after Tinbergen 1959).

Fig. 56. Young male Grant gazelle in a submissive posture before a threatening adult buck. The illustration shows that the behaviors are extreme opposites of each other (after Walther 1965).

limited. This can be achieved by the establishment of permanent territories that need to be defended, primarily during their establishment, by fighting and threat behavior. Later on they are accepted, especially by territorial neighbors, with very little additional fighting. Thus, the main goal of intraspecific aggression, the dividing up of the ecological niche among members of one species, is achieved with a minimum of adverse results for the individual animals.

Territorial behavior not only decreases the number of aggressive encounters as a whole, it also makes possible the alternation between time- and energy-consuming activities, and so it offers an additional advantage. Hence, territories are generally established long before the young are raised, and frequently even before pair formation of the parents has taken place. Hence, the peak of aggression, which coincides with the establishment of the territory, has already passed before sexual and parental behavior appear.

A TERRITORY is defined as a "selectively defended home range," i.e., an area in which the presence of its occupant excludes the simultaneous presence of a same-sex adult conspecific, or less frequently of *all* conspecific rivals (and sometimes even other species; see Section 8.1.5.5). Although all territories have the function of assuring their owners(s) of certain spatial limits within their ecological niche, there are substantial differences among species according to their life-styles and needs for food. The important functions of a territory are as follows:

- A territory insures an adequate food supply for its possessor(s).
- It serves as a common meeting place that facilitates pair formation and mating, especially in instances of *Ortsehe*, where the bond of the members of a pair is to the territory rather than to the mate (see Section 8.2.5).

- ☐ It enables an animal to become very familiar with an area. It knows escape routes and can quickly reach safety and protection against predators.
- ☐ It facilitates finding of food.
- ☐ It reduces interruptions of reproductive activities by conspecifics.

In most cases a territory fulfills several of these functions simultaneously. Hence, it is not always possible to say in a specific instance which were the initial selective advantages under which the specific type of territory of a species evolved, and which of the current functions are secondary effects.

Territory ownership occurs in all classes of vertebrates and is even found in some invertebrates, e.g., brittle stars, crabs, spiders, and insects—especially crickets, dragonflies, praying mantids, and some sawflies and wasps.

Conceptually we must distinguish a TERRITORY from a HOME RANGE. This is an area that an individual, pair, or group habitually occupies but from which conspecifics are not excluded as they are from territories. This is frequently a "neutral" area between two or more territories. Here spatial separation is the result of simple avoidance by animals or groups. It is not always possible, however, to separate these two concepts, since mutual avoidance and active defense can alternate. There are some transitions, especially to temporal territories (see Section 8.1.5.1c).

8.1.5.1. Classification of Territories

The classification of territories that are found in the animal kingdom can be made according to various criteria. We can here present only a few typical examples, among which there are many transitions.

a. Classification According to Function. Most of the previously listed functions of a territory involve the functional systems of feeding and reproduction. Hence, one can distinguish, broadly speaking, among feeding territories, reproductive territories, and territories with duplicate functions, although the dividing line even here is not always easy to draw.

The Feeding and Breeding Territory. This most prevalent kind of territory is used both for feeding and reproduction. It largely meets the needs of parents and young, and its borders are rarely if ever crossed.

The Feeding Territory. Some species (shrikes, robins, and other migratory birds in their wintering grounds) are also territorial outside the breeding season. Such winter territories are the exact opposite of a breeding territory, and they serve only the purpose of securing food for the individual. The same purpose is served by the individual territories of some mammals (e.g., European hamster) and fishes.

The Breeding Territory. Territories can be established exclusively or primarily for raising of young. This is especially true for the breeding territories of colonially seabirds that defend only the immediate area around the nest and search for food on the open sea. Their territory only insures a safe place for raising young. Another function related to reproduction is met by the courting territories of some species of birds: In grouse and ruffs, where there is no permanent pair bonding and the sexes come together only for mating, the males defend their courting arenas against rivals of the same species, which enables them to mate without being disturbed. This same territorial function is also found in some tropical hoofed animals (Grevy zebra, onager, and wild donkey), as well as in the wide-lipped rhinoceros. There seems to be no direct correspondence between the size of the territory and the availability of food.

b. Classification According to Participation in Territorial Possession. *Individual Territories.* Pure feeding and many breeding territories are occupied and defended, as a rule, by only one individual. In species where only one sex cares for the young (hummingbirds, some coral cichlids), the combination of the two functions can have the characteristics of an individual territory.

Pair Territories. In the majority of cases, breeding territories are occupied by both sexes and their offspring. The manner of defense varies. The territory may be defended by both, only by the male, or (but rarely) by the female (button quail, some species of cichlids). An intermediate position between individual and pair territories is occupied by the polygynous cichlid *Lamprologus congolensis* (see Section 8.2.5). In this species individual females maintain their own subterritories within the territory of their male.

Group Territories. In many mammals (monkeys, predators, rodents) and birds (hornbills, some incubating cuckoos, the Australian magpies, and kookaburras) territories are occupied by one group that developed from a family (see Section 8.4.5). Here both sexes, and sometimes the young, participate in their defense. The previously presented general definition does not apply completely to this special variation of a territory since here even members of the same sex may live in the territory. However, this tolerance is restricted to the members of the same group.

c. Temporally Organized Territories. Many species are only territorial in certain seasons. This is especially the case in animals where the territory fulfills some functions during reproduction that is seasonally limited in most species that live in moderate climatic zones (see Section 8.2.2.4). In contrast, "winter territories" exist only outside the breeding season. Only a few species (e.g., badger, fox) live all year in their territories. In the tropics, permanent

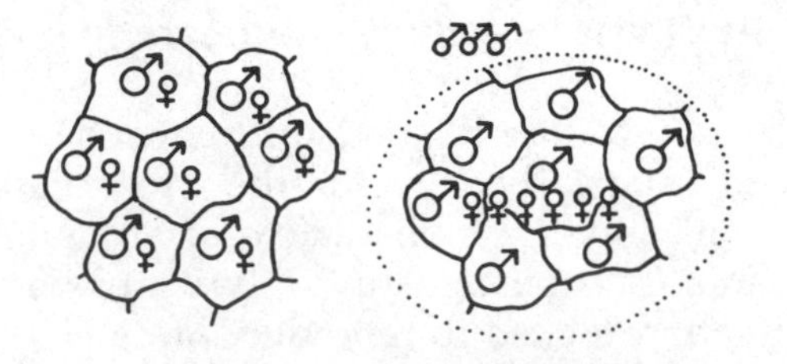

Fig. 57. A schematic representation of two territorial types that occur in gazelles and antelopes: on the left, mating and feeding territories are combined, and on the right are pure mating territories. The dotted line indicates the border of the mating site. The small ♂♂♂ symbols outside the dotted line indicate nonbreeding males (after Klingel 1975).

territories are more frequent and have been described especially for many bird and mammalian species.

In contrast, there are territories of extremely short duration. In the migrating plains population of the African wildebeest, the males establish mating territories within a matter of minutes. They maintain them for several hours or days and leave them as soon as the herd moves on. Migrating birds can also exhibit temporary territories on the way to and from the wintering grounds. In the territorial behavior of the dragonfly *Aeschna cyanea*, the temporal character of the territory is still more pronounced. The males remain only from 10 to 40 minutes at a mating place. There they try to mate with females and drive off other males in aerial combat. Toward the end of this period, their readiness to fight decreases, and the "territory owner" is driven off by a newly arrived male or he leaves the scene without a fight. In mammals, temporal territories have been described in cheetahs and free-living domestic cats. In these species several individuals or several groups adhere to a more or less stable schedule in visiting a particular area. Thus, they use the same territory, but they do not meet. Apparently they can determine, by the age of the scent marks deposited by the previous "passerby," whether or not a certain area of the territory is occupied (see Section 8.1.5.4).

8.1.5.2. Internal Structure

In many species, especially in mammals, specific activities take place in specific locations within the territory. These fixed points are connected by paths with one another and include sleeping area, feeding places, urine marks and dung heaps, drinking and bathing places, wallows, rubbing trees, lookout mounds, and escape and hiding places. An especially important central point (provided it is present in the territory in question) is the den or nest.

Very large territories may have the character of a "path system with goal points," where the area between them is rarely if ever defended, i.e., is not actually a part of the territory as such.

8.1.5.3. Territorial Boundaries

The size of a territory is to some degree species-specific and depends on its function and on the body size and food requirements of the species. Hence, breeding and courting territories are smaller than food territories. Larger species usually have larger territories than smaller ones, and predators have larger territories than plant-eaters. With differing styles of life, territory size may show considerable differences even among closely related species. The largest territories among plant-eaters are known for Grevy's zebra and for wild asses; and among predators, e.g., tiger and lion, they may encompass an area of several hundred square kilometers.

Within the species-specific variability, territory size may fluctuate in the individual case. Older and experienced males often have larger territories than younger ones, and in optimal biotopes, territories may be smaller than under

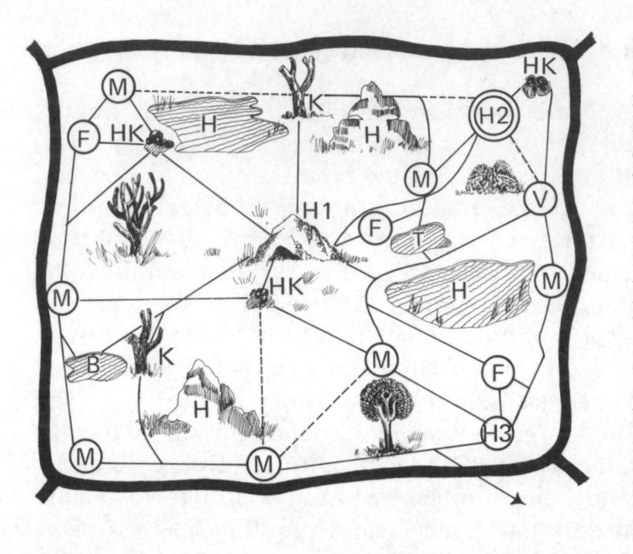

Fig. 58. Schematic representation of the internal arrangement of a mammalian territory. The heavy line signifies the territorial boundary, the thin lines indicate paths that connect the various fixed points, and the dotted line shows less frequently used paths: H1 is the first-order home, i.e., the normally used den; H2 is the second-order home; H3 the third-order home. H is an obstacle; HK is a defecation and urination site; K is a comfort station (e.g., rubbing tree); B is a bathing site; F is a feeding area; V is a food cache; T is a drinking place; M indicates marking sites (after Hediger 1954).

marginal conditions. An especially strong influence on territory size is exerted by population density, which depends on a number of factors such as food availability and predation pressure. The more densely populated an area, of necessity the smaller the individual territories become. Such a limitation is, however, possible only down to a certain minimum size, which has developed during the course of evolution in a way as to assure the specific function of the territory. The maintenance of the minimum size is the result of the territory owner's readiness to fight, which decreases from the center to the periphery of the territory, where the tendency to flee increases in the same measure. For this reason, the peripheral areas of a territory are less intensively defended than the center, and they are most frequently lost. In the center, the readiness to fight is so great that the animal usually prevails in these conditions. Territories have been compared with rubber disks, which are harder to compress the smaller they are. There are, however, some species where the most severe fighting occurs at the periphery, e.g., Thompson gazelles and others, but their territories do not have homes or dens that are usually in the center of the territories.

8.1.5.4. Marking Behavior

So that a territory can fulfill its specific functions, it must be recognizable as an already occupied area. This can be accomplished in several ways and can involve several sense modalities.

a. Visual Marking of Territory. A territory can be recognized by others by the conspicuous display of the owner. This involves the use of prominent landmarks or the exhibition of specific movements or body postures as conspicuous signals. The brilliantly colored males of the Demoiselle dragonfly (*Calopteryx*) fly conspicuously around the borders of their territories, taking off from various specific resting places. Male hartebeest stand clearly visible on termite hills and other high places in their territory.

b. Acoustic Marking of Territory. Animals that can produce sounds often use vocalizations to mark their territory. The best-known example is the song of birds, which is most frequently heard when the territory is being established and which keeps new arrivals out of an already occupied area. The far-carrying vocalizations of seals, monkeys and apes (howler monkeys, gibbons, and orangutans), frogs, and lizards (geckos) can identify a territory to their conspecifics. Even in some species of fishes, there is an acoustic marking of territories.

Sometimes, sounds other than voice are used, e.g., the drumming of woodpeckers or the rattling of snipes, produced by the outer tail feathers, which vibrate in the airstream. It is possible that the knocking and vibrations of some crabs may also have a territory-marking function, but this has not yet been demonstrated.

c. Olfactory Marking of Territory. Macrosmatic[4] mammals frequently mark their territory with the aid of odor substances. These are substances produced regularly by the animals and whose marking function is secondary (urine, feces, saliva). However, odor substances for marking purposes can also

[4] Animals with a well-developed sense of smell are called MACROSMATIC, while MICROSMATIC species are those with a limited sense of smell. This classification is especially used for mammals, where, in contrast to birds, most species have a well-developed sense of smell. The primates are an exception.

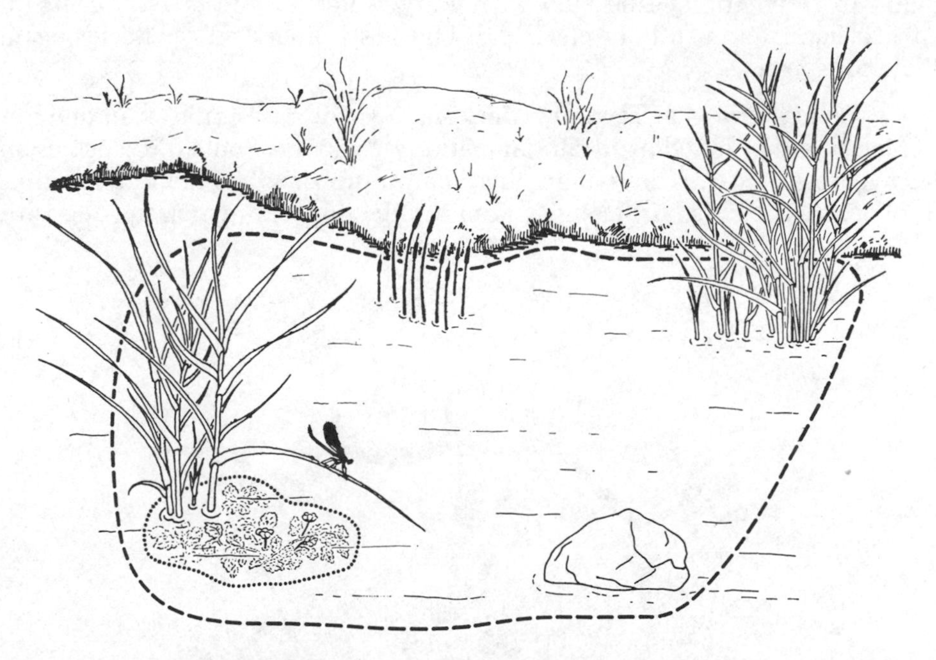

Fig. 59. Territory of a Demoiselle dragonfly male. The territory owner sits, with a conspicuous body posture, on his main perch in the vicinity of the egg deposit site (dotted line). Periodically he flies around the same paths along the territorial boundaries or circles within the territory (after Heymer 1973).

be produced in special glands. Hippopotamuses mark by spraying of dung; several marsupials and rodents use saliva; and urine marking is found among others in dog- and catlike predators, rhinoceroses, prosimians, and some rodents. Special scent glands are frequently located near the anus (anal glands in mustelids) or on the head (preorbital glands of many hoofed animals), but also on other parts of the body. Some species have several kinds of scent glands. Depending on their participation in territorial defense, scent glands are present in both sexes or in males only. Frequently they are fully functional only during the breeding season. Their secretions are deposited preferentially on exposed places (tips of branches and grasses). In some mammals odor marks have a dual function. Not only are they used to mark the territory but they are also fixed landmarks that help the owner(s) to orient themselves. This is especially important in very large territories.

Other than in mammals, olfactory marking of territories has been demonstrated in only a few species. Bumblebee males secrete odor substances from their mandibular glands, with which they make their courtship territories. The solitary bumblebeelike *Euglossini* also mark with odors, but they use substances that they have collected from flowers.

d. Electric Marking of Territory. A number of fish species have the ability to produce discharges that are produced by special electric organs consisting of modified muscle tissue. The electric currents they produce are used in the capture of prey, in defense, and—by distortions in the surrounding electric field—in orientation, especially in muddy waters. At least one group, the African tapir fish use their electric discharges to mark off territories against their conspecifics.

e. Combinations of Territory Marking. In some cases more than one sense modality is stimulated by territorial marking behavior. Some bird species sing from exposed perches or in flight, thus combining visual with auditory signals. In olfactory marking, visual signals can also be involved, when the deposition

Fig. 60. Oribi antelope depositing a scent mark on the tip of a blade of grass (after Hediger 1954).

of urine, feces, or glandular secretions is accompanied by conspicuous body postures or movements.

f. Comparative Considerations. Each sense modality used for territorial marking has specific advantages that are inherent in the characteristics of each modality as well as in the way of life of the animals that use them (e.g., a diurnal or nocturnal way of life, habitat open or with dense vegetation). To avoid damage in aggressive encounters, odor substances offer the greatest advantages, since they also fulfill their function in the absence of the territory owner. Hence, they are less dangerous to him than any visual or auditory marking could possibly be. With respect to visual marking, only one parallel is known: In the crab *Ocypode saratan* the territory, which functions as a home and courting and mating place, is marked by a sand pyramid built by the crab. It is a permanent mark that does not require the presence of its builder.

Odors also last longer as effective markings than any other. They are especially suitable for the marking of temporally overlapping territories (see Section 8.1.5.1). Acoustic signals have the advantage of wide individual variation and offer special opportunities for the individual recognition of neighbors. This greatly reduces fights between established neighbors, which are then reserved more for newcomers who are more of a threat to the territory owner. The advantage of visual signals for signaling is that they can be immediately localized. Hence, they are especially useful over short distances.

In some cases, marking behavior can fulfill two functions at once: It repels other males and also attracts conspecific females. Such a dual function is found especially in the song of songbirds. While its female-attracting function has not been finally proven, there are many indications for it. In many birds, especially in migratory species, the males occupy territories in the spring and sing frequently and for long periods until they have paired off with a female. After that, their singing declines noticeably. Even after pair formation, the attraction by song may continue. In many species, paired males sing especially after they have lost contact with their female. Here, the song also seems to be involved in pair formation (see Section 8.2.3). In many batrachians, e.g., in most toads, the call note of the males seems to attract females and keep other males away.

For purposes of clarity, we discussed threat and marking behavior in separate chapters. However, it should be emphasized that all marking is at the same time a threat, and that the behavior patterns involved either overlap or are in part identical, and that marking behavior may be considered a low-intensity form of threat. The difference is primarily in the situation: Marking is largely nondirected and occurs also in the absence of a rival, whereas threats are oriented directly at him. However, this difference is so minor that some authors reserve the term TERRITORY MARKING only for those cases where actual "marks" are deposited, and which are also effective in the absence of the territory owner (scent marks, sand pyramids of the crab *Ocypode saratan*), while they consider all behavior patterns that coincide with territorial defense as threat behavior.

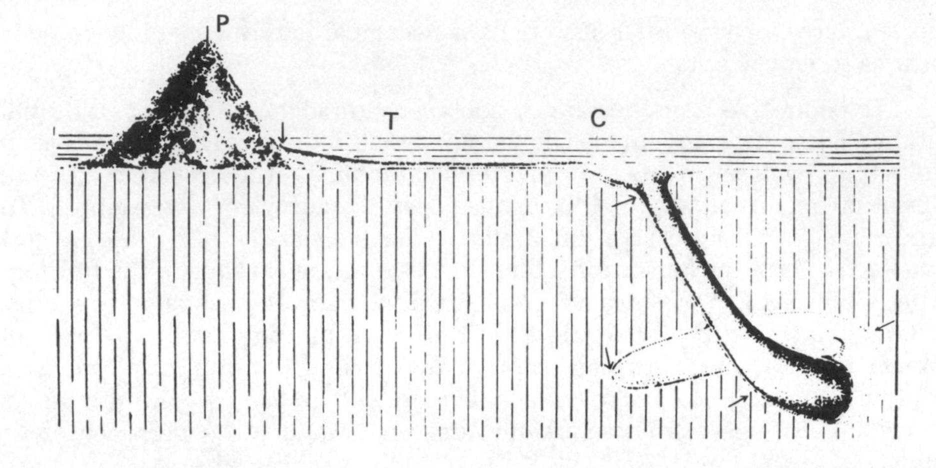

Fig. 61. Mound of a male crab, consisting of a pyramid (P), path (T), "courtyard" (C), and spiral cave. The cave is the home of the crab. The pyramid indicates the presence of the territory owner to potential trespassers, even in the owner's absence (after Linsenmair 1967).

The numerous facts that were cited in this discussion of territorial marking should not obscure the fact the most of them are mere observations. Proof of a repelling function is available only in a few instances. Experimental investigations were carried out in the North American red-winged blackbird. This species breeds in loosely organized colonies within which the males maintain small territories. Here they sing with wings spread far apart, displaying their conspicuous wing patch. If the males are silenced by cutting the nerves leading to the larynx, or if their wing patch is painted over, they lose their territories, or at least they suffer many more violations of their territorial borders.

From this review we can see that territorial behavior as such does not exist, but that instead we are dealing with a very complex phenomenon that adapts the size and function of the territory as well as the duration and manner of occupancy to the biological needs of a particular species. This has come about as a result of the appropriate selection pressures during the course of evolution. Great differences between territories are evident, especially with respect to feeding.

8.1.5.5. Interspecific Territoriality

Territories exist to keep away immediate competitors; hence, they are usually defended only against conspecifics (see also Section 8.1.5). However, when two species living in the same area make very similar ecological demands, thus coming into direct competition, it is of advantage if the members of the other species are also excluded from the territory. Such examples of interspecific territoriality are known among woodpeckers, hummingbirds, nectar birds, reed warblers, shrikes, wheatears, and crows. The species in this category frequently are very similar in appearance, in their vocalizations, and in certain behavior patterns, which all constitute a kind of "marking" that is understood by all, and which have the function of interspecific releasers (see Section 3.5). Some ant species also defend the areas in which they "maintain" aphids not only against conspecifics but against other species as well.

8.1.6. Individual Distance

H. Hediger distinguished between contact and distance animals with respect to the spatial relationships between species members. Contact animals tolerate body contact, while distance animals keep away from each other and attack any conspecific that oversteps an "individual distance," or that will not move out of the way.

Both types are apparently randomly distributed among various groups of related species. They have been investigated more closely primarily among vertebrates. Contact animals include pigs, hippos, most rodents, many monkeys and apes and prosimians, most parrots, white-eyes, mousebirds, turtles, some lizards, salamanders, eels, moray eels, and most catfish. Examples of distance animals are ruminants, flamingos, gulls, most raptors, pike, trout, and salmon. Yet it is not possible to find a correlation with the structure of the body surface (porcupines and hedgehogs are contact animals) or with the need for gregariousness, since many colony breeders and herd animals are true distance animals.

There are all kinds of transitions between these two types. In some species, body contact is merely tolerated, in others—especially during resting periods—it is actively sought, with as much body surface touching as possible. Furthermore, contact behavior can be restricted to specific situations (danger, bad weather), to specific individuals (partners of a pair, parents and young), certain age groups (young animals), or to one of the two sexes. In sexually dimorphic zebra finches, the females, but not the males, tolerate contact. In two situations even extreme distance animals will touch each other: during mating and when caring for young.

Even within distance animals, there are many differences. Thus, the individual distance is in some cases quite similar for all members of the species, as in swallows or starlings sitting on an electric wire, while in other species individual or sexual differences exist. Male chaffinches keep a distance of 18–25 cm, females a distance of 7–12 cm.

True contact behavior probably evolved as protection against external danger. The closed circular formations of musk-oxen are a defense against predators (see Section 8.4.1.1). The mass congregations of Emperor penguins, or the clusters of resting swallows and swifts, protect them against heat loss. In other species a direct correlation with environmental factors seems not to exist. An extreme example in this respect is found in the Australian wood swallows, who form clusters and sit close to one another even in temperatures of 50°C and more. It is possible to assume that this contact behavior, which

Fig. 62. Individual distances between perching swallows.

originally may have been an adaptation against heat loss, secondarily became important in pair and flock formation.

On the other hand, the existence of an individual distance may be important to assure a "private" zone for an animal, e.g., a certain amount of food in grazing or other animals that feed in groups, where it reduces interference. Therefore, individual distances have frequently been considered as moving territories, and possibly as the phylogenetic precursor of territories, since they should be automatically identified with a specific area once species became settled. In many discussions, the questions about a conceptual distinction between territory and individual distance have been attempted to establish criteria for these two categories. Space is certainly not a suitable criterion, since the smallest territories (with reference to body size), which are found in the breeding territories of colony-nesting seabirds, correspond to the pecking distance, which is determined by the reach of their bills. The location of a particular place is also not a good criterion, since moving territories also exist. Thus, in quail and some ducks, the males defend a certain area around the females at the beginning of the breeding season. Finally, the already mentioned temporal territories hold an intermediate position, since they do not require a complete possession of space, but instead they constitute an enlarged and temporally extended individual distance by means of odor markings.

However, there are also clear differences that do not depend on the presence or absence of a defensive reaction at a particular place. Thus, when defending itself in distance behavior, the animal fights only when its individual distance is violated, whereas in territorial disputes other objects, e.g., the nest and young, are defended. Beyond that, the defense of a territory is much more closely tied to reproduction than is the maintenance of an individual distance. Hence, it seems prudent to maintain these distinctions despite a certain amount of overlap.

8.1.7. Ritualized Fights

Threat, territorial, and distance behaviors fulfill in many situations the same function as actual fighting behavior, but they cannot serve as a complete substitute in all instances. Especially when territories are established, when a rank order is being established (see Section 8.4.6.1), or in competition for a sexual partner, actual fights can occur. However, since these fights are meant not to injure or kill but merely to displace rivals, there sometimes occur behavior patterns in intraspecific encounters that are not found in fights with other species. This is especially true for species that possess dangerous weapons. Thus, poisonous snakes fight by a sideways pushing of one another with their heads, and many lizards press their heads together in a test of strength. In both instances, the sharp or poisonous teeth are not used. Some cichlids fight with tail beating, which propels pressure waves against the head of the opponent, or they grasp each other's lips and engage in a push–pull contest, but they do not damage each other. Many hoofed animals are very dangerous to predators that they can kick, but in rival fights they do not use their hooves. Instead, the animals fight with horns or antlers whose shape is usually not designed to cause serious injuries. In other instances the style of fighting is similar, but the result is different: The initial thrust with the horns leads to a crossing and locking of horns in a push–pull fashion (see Figure 63c), i.e., a harmless form of fighting. A nonspecies opponent that is without the proper horns with which to parry the thrust can be easily gored.

These fights, which frequently consist of fixed behavior sequences, i.e., which follow certain "rules," are called RITUALIZED FIGHTS. The strength of the selection pressure that underlies the development of ritualized fighting can be appreciated when one considers that they require more energy and time than the damaging and killing fighting patterns used between species.

8.1.8. Damaging Fights and Killing of Conspecifics

True damaging fights are rare in comparison to other forms of intraspecific aggression. Many species do not have weapons with which to seriously injure a rival; others usually do not use them when fighting rivals (see above). However, even those species that actually use their potentially harmful

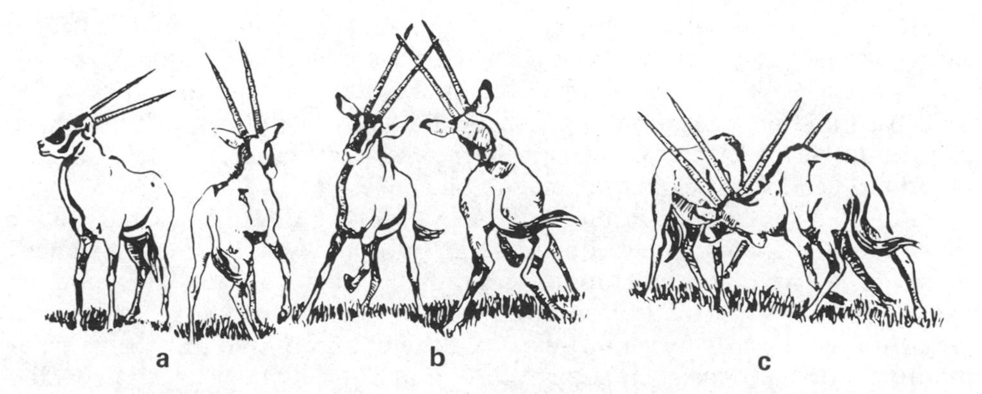

Fig. 63. Fighting oryx antelopes: threat with raised head (a); clashing of the upper third of the horns (b); forehead pushing (c) (after Walther 1958).

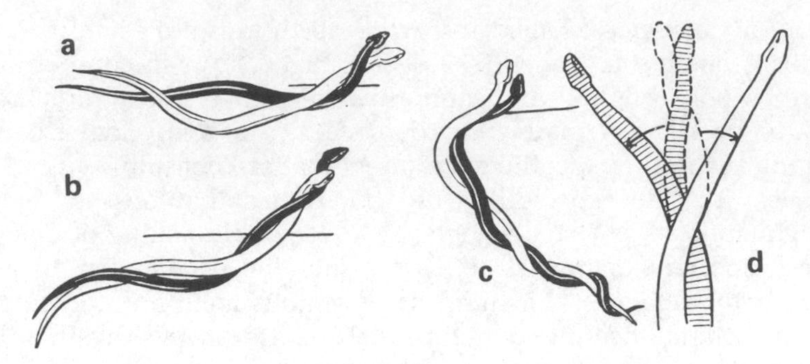

Fig. 64. Ritualized fight of two snakes *(Vipera berus)*. The animals wind around each other for the full length of their bodies and press their heads together, which eventually slip apart. In (d) the slipping apart is shown schematically. In this test of strength the snakes do not use their poisonous fangs (after Thomas 1955).

weapons can reduce the aggression of a superior opponent with appeasement gestures (see Section 8.1.4), or they are protected against them by special protective features of their hide or skin. Thus, the bites of fighting hippo males or sea lions rarely penetrate the thick layer of fat below the skin, and in male lions the most vulnerable parts of the body are protected by a heavy mane from the bites and claws of their rivals. These adaptations, which contribute to the prevention of serious damage that could be incurred in intraspecific fights, support the hypothesis that intraspecific aggression is meant to displace rather than kill a rival (see Section 8.1.2). In view of such considerations, ethologists were for decades of the opinion that killing of conspecifics hardly occurred at all under natural conditions. It was thought that fights resulting in death that were observed in captivity were due to the lack of opportunity for the lower-ranking or defeated animal to withdraw to safety. This view no longer prevails without some reservations: The observations of wild animals killing conspecifics in the wild are increasing, while on the other hand, new ideas about the nature of natural selection (see Section 8.4.3.2) have provided even for these cases explanations that suggest a biological function under specific circumstances.

Killing of conspecifics may affect adults or young. In the first instance, there may be situations in which certain environmental conditions "require the sacrifice" of some individuals. This may reduce an overutilization of the food base and hence prevent a threat to the entire population. Under these conditions, the exclusion of conspecifics as a "last resort" would limit the population and thus be of ecological significance.

Fights that result in death have been reported in lions, hyenas, musk oxen, various rodents, and, among birds, in gulls and loons. In most instances these are species that live in groups and those killed are usually strangers (see Section 8.4.4.2). Apparently, inhibiting mechanisms do not operate, or operate very little, in fights between members of different groups. In this respect, the readiness to fight members from other groups is similar to interspecific aggression (see Section 8.1.1).

The limitation of inhibitions to members of one's own group is biologically meaningful, since members of different groups are usually not related to one another; hence, they do not contribute as a rule to the increase of their part of the gene pool in the total population, according to the "fitness" calculations outlined in Section 8.4.3.2. (Scientifically it is of interest that this limitation is probably the reason why the killing of conspecifics has been recognized so late as a natural behavioral element in some species. Earlier investigations dealt more with behavior *within* social groups, where the inhibition against killing group members is very strong.)

Killing of young (infanticide) occurs in two different situations. It can be done by the mother and by strange males. The former instance can occur if the young are ill or retarded in their development, but healthy young may also be killed when environmental conditions do not permit raising of young without disturbance. Even when the simultaneous raising of young is possible, the last one born may be sacrificed in favor of older young in which more parental care has already been invested. Killing of young in the wild is known in single instances in the wild pig, red fox, European stork, and several species of songbird. In captivity it occurs more frequently as a result of disturbances and social stress. In many cases the young are not only killed but eaten as well.

Infanticide by strange males has an entirely different function from that due to unfavorable environmental conditions. It has been observed in species that form harems (see Section 8.2.4), especially in lions and among primates such as the Indian langur. If in the latter a male has vanquished a harem owner and taken over the females, he will kill all their helpless young without, however, eating them. In this way, females who during the lactation period do not come into estrus as long as they nurse young will soon come into estrus again. Apparently this behavior maximizes the chances of passing on the genes of the males to the next generation (see Section 8.4.3.2). Thus, the killing of the previous harem owner's young, with whom the new male is not, as a rule, closely related, leads to an earlier opportunity to produce his own offspring.

8.1.9. Motivational Aspects

The wide distribution and important biological function of intraspecific aggression raises the question of how the aggressive motivation of an animal is controlled. Several subsidiary questions arise: Are there indications of an innate basis of aggression? How important are environmental factors? What are the integrating mechanisms in the organism? Is there a specific "aggressive drive"?

These questions, which are especially important for an understanding of human behavior, are by no means answered to date, and they are still quite controversial. Still, there are numerous results that can contribute to a clarification of these matters.

8.1.9.1. The Species-Specific Extent of Intraspecific Aggression

An important indication for the existence of a genetic basis is the fact that the extent of intraspecific aggression within an individual species operates within certain limits and often changes with the seasons (see Section 8.1.9.4); hence, it differs greatly from one species to the next. Even among closely related species there are obvious differences. Among gannets, a colony-nesting seabird that nests on steep cliffs, aggression is much higher than in the Cape gannet, which nests on open ground. These differences are correlated with the nesting practices of each species. Gannets nest on cliffs where there is a need to defend suitable ledges against all others. The Cape gannet, on the other hand, has adequate room for nests, hence is less aggressive. It would even be a disadvantage because those birds that nest in the center of the colony have to pass through breeding territories of other colony members on their way to a suitable launching site at the shore. If Cape gannets were as aggressive as gannets, there would be continuous conflicts even between established pairs. Similar examples in many other species show that each has evolved the ecologically most appropriate amount of intraspecific aggression during the course of evolution.

8.1.9.2. Artificial Selection

All of these species differences and ecological adaptations suggest the hypothesis that the average level of intraspecific aggression is determined by the genome. It can be held constant by natural selection (see Chapter 10), which controls a readiness to fight that is too high or too low by reducing the reproductive success of individuals with these extreme characteristics. In the same way, changing environmental conditions can lead to new adaptations.

That such a model can have validity is shown when artificial selection is introduced in place of the more slowly operating natural selection. Thus, in many species it has been possible to select the most aggressive individuals and to produce, during the course of domestication, increased aggressive behavior in the Siamese fighting fish, in fighting bulls used in the ring, and in chickens (see Section 11.3). In house mice, various inbred strains with varying degrees of fighting motivation were developed by selective breeding. Here the genetic basis of strain differences was confirmed by genetic experiments: If one crosses mice of the various strains with one another, the aggressivity of the F_1 generation is intermediate between that of the parent strains (see Section 9.2).

8.1.9.3. Effects of Experience

The presence of a genetic basis does not, however, permit the conclusion that aggression is inborn as such, i.e., that environmental influences play no role in the development of interspecific aggression. In fact, it was found that in some instances a very profound influence of social experience could be demonstrated. Long-lasting influences were especially pronounced when the experience took place early in life. Male mice that were gently stroked as young were especially peaceful as adults. On the other hand, mice that had to fight for food early in life were very aggressive later on and fought about food even when they were not hungry. Generally speaking, success in fights can increase aggression later in life, whereas frequent losses decrease the readiness to fight. These effects are more obvious the earlier in life the relevant experiences have taken place. Young rhesus monkeys two to three months old, who were raised by mothers who grew up without mothers themselves and who were very aggressive as a result, were very aggressive as adults, and showed twice as many aggressive behavior patterns as conspecifics raised by nonaggressive mothers. Similar obersvations exist in other species. Aggression can also be increased by pain (electric shock), but the effects vary depending on the shock intensity.

Overall, the literature shows that for species-specific aggression the genome defines the limits or range of its expression (see Section 6.1.1), and that its expression in the individual is determined by experience within this range. Here too, then, we find an intimate relationship between genome and the environment. This has not always been sufficiently considered in the history of ethology. Repeatedly the innate and environmental sides of aggression have been seen as actual alternatives, which has led to vigorous controversies among several ethological schools.

8.1.9.4. Integrating Mechanisms

With respect to integrating mechanisms, intraspecific aggression cannot be distinguished from other functional systems. Experimental evidence exists for hormonal as well as central nervous system control of behavior. Hormonal control is of advantage in all instances where the level of aggression depends on long-term, i.e., seasonal, fluctuations (see Section 5.2). In such instances it reaches a maximum when the demands for aggressive behavior are greatest, as during the establishment of territories and in combat with rivals. Thus, the red-bellied model that, in the spring, releases vigorous attacks in a male stickleback is not responded to at all during the fall. Because of the temporal parallel between the two behavior systems, a control involving the sex hormones seems to be biologically most meaingful. Castration experiments and hormone replacement treatments have shown that such a connection does in fact exist.

However, the picture of hormone involvement in the control of aggressive behavior is quite complex: Androgens generally increase aggression, whereas the role of female hormones is not clear. In addition, the hormones of the adrenal glands and the gonadotropic hormones of the pituitary gland can affect the aggressive behavior of an animal (see Section 5.2.2.4) in various ways, but here, too, the actual relationship has not been clarified.

There are many indications that androgens, especially testosterone, influence primarily aggressive behavior that is directly involved in reproductive activities, e.g., territorial fights, fights about a female or for nesting material. These statements are primarily based on results obtained from males, but apparently they are also valid for females. In the North American white-crowned sparrow, where the female actively participates in territorial defense, an increased testosterone level was measure in the female. All other hormones, especially pituitary hormones, whose effects can generally also be demonstrated in females, seem to control conflicts of a more general nature.

8.1.9.5. Appetence for Aggressive Behavior

Our discussion about the control of aggressive motivation culminates in a discussion about the presence and characteristic of an independent "aggressive drive" (see Section 2.6 for a discussion of the concept DRIVE). Basically, there are two possibilities: Either the temporally fluctuating readiness of an animal is merely the result of environmentally obtained information (e.g., seeing a rival, territorial stimuli) or, in addition, there are spontaneous factors that are independent of the environment. This alternative between "reactive" and "active" is independent of the question of a genetic basis of aggression, since it does not have to manifest itself only in the form of a spontaneous production of excitatory potential but also on specific reaction to external stimuli, which varies from one species to the next.

Results about the question of the possible spontaneity (see Section 2.5) of aggression do not permit clear answers to date. Pure vacuum aggression has not been demonstrated as yet; however, true vacuum activities are hard to demonstrate in any case (see Section 2.2). Nevertheless, there are many indications that argue for a damming up of a readiness to fight. In some cichlid fish and in many pheasant species, males attack and even kill their females when they are kept only in pairs, i.e., when they do not have an opportunity to fight with other males. On the other hand, if they can fight with other males, even if only through a glass partition or wire mesh, the females remain largely unmolested, and successful mating can take place.

Such an increase in the readiness to fight in males who have no rivals to fight with argues for a damming up of aggressive motivation, but alone it does not prove that this damming up is completely endogenous. At least in cichlids that form pairs, the sexes look very similar. Hence, it is possible that subthreshold stimuli, which gradually increase the males' readiness to fight, emanate from females (see Section 2.2). Investigations on other cichlid species point in this direction. These show that by constant exposure to adult conspecifics, or models of these, aggression increases. In pheasant such an induced increase is less likely since both sexes are sexually dimorphic.

Keeping an animal in complete isolation also leads to an increase in fighting readiness. In the hermit crab *Pagarus samuelis*, the increase is continuous so that aggressive behavior patterns increase at first at a lower, and later at a higher, releasing threshold. Finally, investigations on chicks and mice have shown that rearing in complete isolation results in especially aggressive individuals.

In contrast to the method of keeping animals in pairs, keeping them in isolation has the advantage that all influences by stimuli coming from conspecifics are certainly excluded. However, this advantage is offset by a decisive disadvantage: Singly kept animals often show various developmental abnormalities, which manifest themselves in an increasing sensitivity toward *all* environmental stimuli. For this reason it cannot be determined here whether the increased reactions to aggression-releasing stimuli depend directly on an increase in aggressive motivation or whether they are merely an indication of a more general sensitivity of the organism. Overall, then, the available evidence to date does not allow us to draw conclusions about the accumulation of aggressive motivation, i.e., for a spontaneous component of aggression.

The same is also true for a specific appetence for fighting. Its existence can be tested experimentally by operant conditioning (see Section 7.4.3), where the opportunity to fight is the "reward." Some birds and fishes can be trained to seek out a specific place if they can fight there with a rival, or if they are presented there with a threatening model or with their own mirror image as a rival. Hamsters and mice chose the runway in a T-maze that led to a rival that they could fight. Mice that had been separated after a short rival fight even crossed electrified grids to return to their opponent. In the latter case the voltage that was tolerated even provided a quantitative measure for the readiness to fight at a particular time.

Such results can be considered important indicators for the existence of a specific appetence for fighting behavior. However, these results still are no proof for its spontaneous nature, since even in animals raised in complete isolation it cannot be excluded that the animal is stimulated aggressively by "neutral" stimuli from the environment, which it then associates with earlier aggressive encounters. This, however, would again indicate a possible exogenous source for the control of aggressive motivation.

8.1.9.6. The Biological Significance of an "Aggressive Drive"

Although the final evidence for the existence of a true appetence for fighting has not yet been supplied, the sheer number of already available indications for it justifies a discussion of the biological significance of such an appetence. The function of such a drive system would be to "insure" the availability of behavior patterns that are necessary for the preservation of the individual or the species (see Section 2.8).

In the functional systems of food acquisition and the need for reproduction, such a drive system is obvious. With respect to agression, it is not so apparent since its biological function could also be met if the behavior patterns were available only when needed; hence, they would occur only in the presence of the appropriate stimulus situation. A spontaneous search for a situation that releases fighting would actually be a disadvantage (energy waste, danger) for several reasons.

Despite such statements that argue against an independent motivation for aggression, there are situations in which a specific appetence can be of value. This could apply to marking behavior, since territories can fulfill their function of reducing fights (see Section 8.1.5) only when they can be recognized as such by a potential rival who appears on the scene, i.e., before any meeting takes place.

A second example refers to a situation in which the external stimuli are present, but in which the appropriate reaction to them is changed. This we find in species that exhibit intensive care of young in the phase of development in which the bond between adults and young dissolves. Investigations in several primate species have shown that the initiative for this process can be taken by either side. Such a sudden change from positive to negative social relationships leads to the postulation of an increase in aggressive motivation, which is responsible for the difference in the reaction to previously positive social stimuli. This is especially obvious when the adult male changes from his role as father to that of rival. A veritable burst of increasing aggressivity can be seen in many songbirds at this time, where it possibly contributes an expansion of a species' range. At least it seems possible that such changes in the motivation systems have an endogenous, i.e, spontaneous, basis.

8.1.9.7. Conclusions

The results reported in this chapter show two things: Investigations dealing with intraspecific aggression in animals are beset by significant methodological differences because objective measures for aggressive behavior patterns are hard to develop, since aggressive motivation cannot be measured directly (see Section 2.7) but only with the appropriate behavior patterns. Furthermore, the influence of possibly interfering factors cannot always be determined or excluded. More than in any other functional system, the proof depends on the appropriate methodology.

The form of the aggressive behavior patterns, their biological function, the relative contribution of genetic and experiential control, the possibility of damming up of the drive, the influence of hormones, and many other parameters differ from one species to the next. Hence, a warning about generalization across species is in order here more than in all other functional systems. Results of studies on one species should not be applied to others, nor is it warranted—as has occurred several times—to draw conclusions to refute findings obtained in one species by those obtained in another.

8.2. Sexual Behavior

8.2.1. Biological Significance

Sexuality is one of the basic phenomena of life. With the exception of some lower organisms, sexual differentiation and sexual reproduction are known in all living organisms. There are species that reproduce asexually through one or more generations, i.e., by cell division or budding. However, even here one finds sexually reproducing generations in regular or irregular intervals.

The biological significance of sexuality is the same for all organisms: During the formation of gametes, cells divide, during which the chromosome number is reduced to half so that the gametes, sperm and ova, receive only half as many chromosomes as the other cells in the body. During fertilization, there is a recombination of genes derived from the father's and mother's genetic material in the newly formed zygote. Sexuality thus provides the basis for the variability of organisms, which in turn is the basis for their adaptability to the environment, and hence the further development of all living things.

The most important event in this process is fertilization, the combination of sperm with the ovum. It can be preceded by copulation. During fertilization one sperm penetrates an ovum. By FERTILIZATION is meant the penetration of a sperm into the ovum. COPULATION refers to the joining of the male and female sexual organs. The term MATING is also used sometimes, but a distinction needs to be made in the case of those species of fish that have a gonopodium that penetrates the female for the deposition of sperm (copulation) and the release of sperm over the eggs once they are deposited by the female (MILTING). These distinctions are often not made in the literature.

The behavior patterns that initiate copulation or mating are called COURTSHIP OF PRECOPULATORY BEHAVIOR. Additional subunits of sexual behavior in the largest sense include PAIR FORMATION and PAIR MAINTENANCE BEHAVIOR. The first-mentioned category includes behavior patterns that lead to a more or less permanent pair bond (see Section 8.2.3); the second category is made up of those behaviors that maintain the bond.

A clear distinction of the three separate categories is not always possible, since in many species courtship does not always lead to copulation or mating, and many of the behavior patterns included under courtship are also used in pair formation or the strengthening of bonds. On the other hand, there are behaviors that are limited to these two categories. Since the boundaries between these concepts are fluid, the term COURTSHIP is not used consistently in the ethological literature. Sometimes it is used only in the narrower sense where it refers to precopulatory behavior; at other times it includes all behavior patterns from initiation of copulation through pair formation and pair maintenance.

For copulation to be successful, a number of morphological, physiological, and behavioral prerequisites must be met. The following apply to behavior:

- The sexual partners must be able to recognize each other by sex and as belonging to the same species.
- They must find one another and join their copulatory organs.
- They must be ready to mate at the same time, i.e., they must be synchronized.
- They must overcome their intraspecific aggression.

8.2.2. Function of Sexual Behavior

It is the function of sexual behavior to insure these preconditions to copulation. Various separate tasks can sometimes be accomplished or controlled by several behavior patterns. As with aggression and care of young, sexual behavior is characterized by many highly developed behavior patterns. Several regularities can be identified in the various categories.

8.2.2.1. Bringing the Sexes Together

In species that do not live permanently in pairs (see Section 8.2.3) or in groups (see Section 8.4), members of the opposite sex sometimes must find each other over great distances in order to mate. The attraction of the partner in the majority of species is the job of the male. Exceptions to this are usually found among insects. Frequently the mechanisms of attraction are identical with those of territorial marking, i.e., the repulsion of rivals of the same sex (see Section 8.1.3). Hence, we will discuss here only such cases as deal primarily or exclusively with the attraction of the sexes. As is the case with territorial marking, several sense modalities can be involved that are effective either alone, in combination, or in succession.

a. Visual Attraction. In many species colored markings, which can be conspicuously displayed, moved (see Section 3.4), or turned "on" or "off," are available to attract a sexual partner. In males of the fiddler crabs which in many species are found along the coasts of tropical and subtropical seashores, one of the pincers is especially enlarged, conspicuously colored in some species, and moved rhythmically. The conspicuousnes of the pincer movements can be further enhanced by accompanying body movements. In tropical frigate birds, the male display is widely visible—the red throat sack develops at the beginning of the breeding season—and in many other species colorful feather patches are unfolded or made visible in some other way. An impressive example is the tail display of the male peacock, with its colorful tail coverts. Visual attraction is also accompanied by the continuous or flashing light signals of lightning bugs. Here the males are attracted by the flightless females.[5] At night they seek out certain, often elevated "flashing posts" and present their abdomen in such a way as to expose the light organs located on the ventral side. In some species abdominal movements further enhance the effect of the signal. The flying males are attracted by these signals during their searching flights. Once above a female, they fold their wings and drop, often with surprising accuracy, next to the female. That the attraction and orientation depend solely on the light signals was demonstrated in experiments with models and mirrors and artificial light sources.

[5] In some lightning bug species, males also possess light organs. Their signals seem to have a synchronizing effect in some species, while in others no influence on females can be detected. Hence, a final explanation about the significance of male light signals cannot be given at this time.

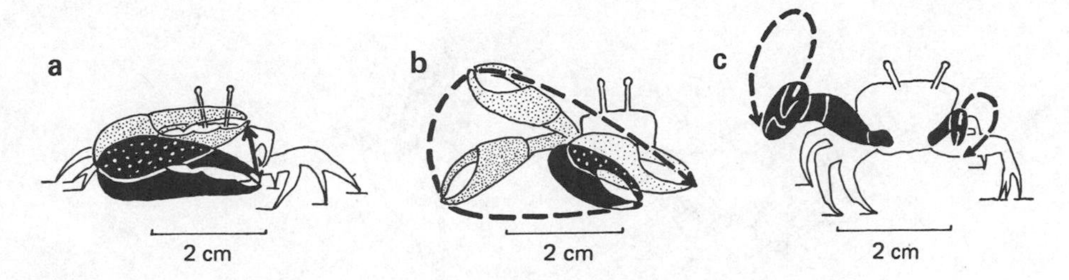

Fig. 65. Pincer movements of fiddler crabs: (a) *Uca rhizophorae* (vertical waving); (b) *Uca annulipes* (lateral waving); (c) *Uca pugilator* (waving with laterally extended pincer) (after Schöne and Schöne 1963).

b. Auditory Attraction. Auditory signals to attract sexual partners are found especially in birds, amphibians, and insects. The attractive function of bird song and the mating calls of frogs have already been described (see Section 8.1.5.3). In many insects (e.g., cicadas, water bugs, grasshoppers, and crickets) females are attracted by male vocalizations. The male grasshopper has a special song to attract females in addition to two other songs.

c. Olfactory Attraction. The attraction of a sexual partner with special odor substances is probably quite widespread in some animal groups, but we have only a few detailed investigations. They refer especially to mammals and insects. Male European hamsters produce an odor substance in a gland located in their flanks which attracts females. Odor signals that carry quite far are produced by females of some nocturnal moths. They produce odor substances in glands located at the posterior end of the abdomen that the males detect with the aid of sensory cells located on their antennae. The sensitivity of these cells is surprising: In the silk moth a few molecules of the pheromone bombykol suffice to stimulate the male's antennae. Thus, with favorable wind conditions, males can be attracted from miles away. In the species *Actias selene*, experimentally displaced males returned over a distance of 46 km to the place where newly hatched females were found.

d. Attraction by Communal Courtship. In many species several males congregate and display behavior patterns that attract females in a kind of communal courtship. This is true for the song of grasshoppers and cicadas, where in many species the males sing together. This is similar to the simultaneous light signals of some tropical lightning bugs or the colorful displays of a group of male carp in North America. The males of some hummingbirds meet to perform a group song, where as many as 60 birds have been observed in one song arena. In the white bird of paradise, two males form a symmetric figure together. They take turns hanging upside down from the branch of a tree. In other species of bird of paradise, several males congregate in communal courtship. The same is found in several South American bird groups, e.g., in manakins and cock-of-the-rock. In all these species there is a pronounced sexual dimorphism, and the males possess a

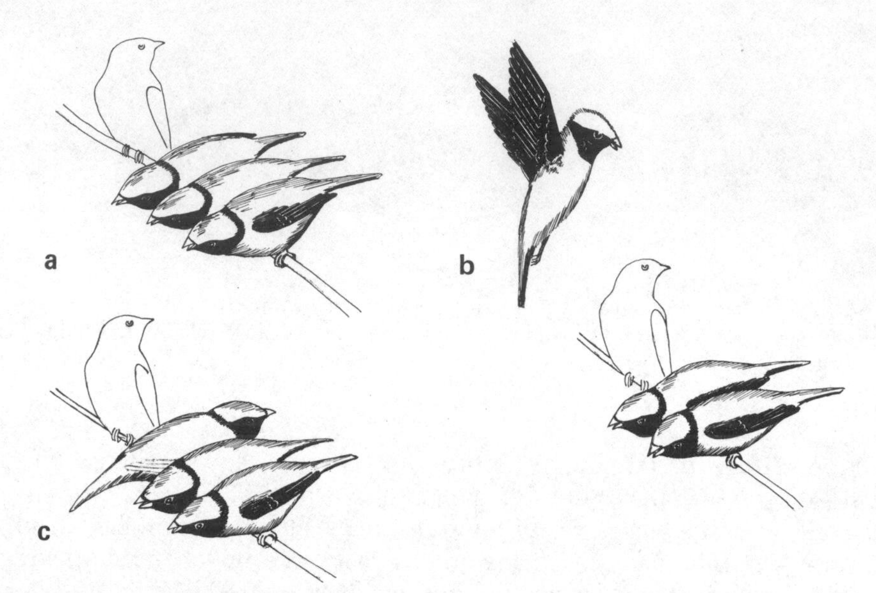

Fig. 66. Three male manakins, *Chiroxiphia caudata*, court one female together (after Sick 1959).

very colorful plumage that serves as a conspicuous releaser. It is probable that these kinds of communal courtship are effective because of the summation of stimuli (see Section 3.6) that have a strong attraction for the females of the species. This developed during the course of evolution under selection pressure that favors conspicuous signals.

The term SOCIAL COURTSHIP has been used to describe the courtship by several males. This is misleading, because in ethological language the term SOCIAL (see Chapter 8) refers to all courtship behavior. Hence, the terms COMMUNAL COURTSHIP or COLLECTIVE COURTSHIP are to be preferred.

8.2.2.2. Overcoming Intraspecific Aggression

The mutual approach involved in the intention to mate or copulate can also result in the activation of aggressive tendencies in the partners. This is especially true for species in which individuals live solitarily, defend individual territories (see Section 8.1.5.1), or maintain an individual distance (see Section 8.1.6). It is an important function of courtship behavior to reduce these aggressive tendencies.

In some species there is much aggressive behavior during and after courtship (see Section 8.2.2.3). Thus, the zigzag dance of the courting stickleback is frequently interrupted by attacks on the female, and in courting zebra finches the courting partners frequently engage in bouts of pecking. The epitome in this respect is found in some spider species, where the frequently smaller male is treated by the female as prey and is killed. This has led to adaptations in which the males of many species have evolved highly specialized signal movements and behavior sequences that identify them to their female as conspecific males and thus distinguish them from potential prey. Still, in some species the male is eaten after or even during copulation.

However, even in animals in which open aggression is absent, it is in many instances possible to detect the phylogenetic origin of courtship behavior patterns that are derived from ambivalent forms that result from interactions between sexual and agonistic components (see Section 4.2.1). As a rule, there is in males a combination of sexual and attack elements; in females, a combination of sexual and escape behavior is more frequent. However, there are also in some spider and in some bird species females that occupy a higher rank order (see Section 8.4.6.1) than males at the beginning of the reproductive season. Hence, many courtship sequences in animals have been considered to be a compromise between sexual, attack, or escape motivation. A well-known example is the zigzag dance of the stickleback: the aggressive motivation predominates in the "zig element," which is directed toward the female, while the "zag element," which points away from the female toward the nest, is sexually motivated. However, such interpretations sometimes resulted in excessive schematization. The widespread function of threat signals as courtship and pair formation behavior has already been discussed (intimidation display; see Section 8.1.3).

In all species where the sexes meet not only for the purpose of copulation but also to maintain a more lasting bond (see Section 8.2.3), the aggression-inhibiting mechasims need to be effective for longer periods of time. In these contexts the same behavior patterns are a component of intention movements to copulation as well as an important means in strengthening the pair bond. Hence, it is especially difficult to distinguish among behaviors involved in pair formation, courtship, and the maintenance of bonds.

Two mechanisms are frequently found that inhibit aggression between members of a pair: the "demonstrative" turning away of "weapons" and the activation of other behavior tendencies that are incompatible with aggression. The first category includes the well-known example of European storks that clack their beaks and bend their heads far over their backs, and the "head flagging" (turning away of their faces) of black-headed gulls, which expose the white backs of their heads to their partners. In this way they avert their black face masks, which play an important role in the threat behavior of this species. In the raven there is a similar behavior in which the powerful beak is turned away from the partner. Many of these behavior patterns resemble appeasement behavior (see Section 8.1.1), since they occur long before an actual fight.

The phylogenetic origins of behavior patterns from other functional systems that inhibit aggression are derived mainly from nest building and care of young. Elements from both systems can be found in the pair-bonding behavior of numerous species. They are especially frequent in many species in the class of birds, e.g., in waxbills (see Section 10.3), or in the crested grebe, where both sexes exchange nesting material during specific phases of their courtship with their bills, or show them to one another with conspicuous bill, head, or body movements.

Fig. 67 (left). Appeasing "looking away" in the raven (after Gwinner 1964).

Fig. 68 (right). Courtship-feeding in bullfinches. The male (right) feeds his female (left) (after Nicolai 1956).

From the functional system, care-of-young behavior patterns involved in the actual feeding have become incorporated into courtship, which includes the components of presenting and giving of food as well as of begging in the receiving partner. As a rule, the male assumes the active role in this courtship feeding. Thus, male terns and seagulls feed their female in the same manner in which the young are fed by their parents: Terns present the captured fish directly with their beaks; gulls, however, regurgitate partially digested food, which is easily picked out of the male's open beak by the female and then eaten. Courtship feeding is also widespread among the various species of parrots. Here the male feeds his female in some species, while in other species there is a mutural exchange of food.

Originally, feeding of the female by the male probably had a physiological function related to nutrition: It supplied the female with food during egg development and incubation. Hence, it is possible that this behavior evolved by selection pressures that favored an optimal supply of food to the female during a time of critical nutritional needs. Secondarily, the function of this behavior was expanded or even changed (see Section 10.6), with the effect of inhibiting aggression in the context of pair formation, a function that the behavior still has today–perhaps even exclusively so.

In connection with this change of function, feeding frequently has only symbolic function: the movements are retained, e.g., in the "billing" of many bird species, but food is no longer passed. In some related groups both possibilities exist. By comparing species, the probable phylogenetic development of this behavior may be traced (see Section 10.3). Thus, the males of many species of fowl still offer food to the female, while in other species they merely peck at the ground or point their bill downward, where in some cases no food is even present.

In the second component of feeding behavior—begging—movements, body postures, and vocalizations have become a part of courtship behavior. The young of many songbirds beg food from their parents by trembling their wings, and the female performs the same movements when soliciting the male for copulation. In some mammals, e.g., the European hamster, the chamois, and the fallow deer, the males produce vocalizations like young in the initial phases of courtship. The elements of juvenile behavior in the repertoire of adult animals are called infantilisms.

8.2.2.3. Postcopulatory Displays

One area of pair bonding behavior in which the aggression-inhibiting effect is also very strong is in postcopulatory displays. These may include in part very complex and highly developed behavior patterns that occur immediately after copulation and that as a rule are not identical with actual courtship behavior.

In the Australian long-tailed grass finch, the male lands close to the female immediately after copulation, bends the anterior part of its body downward, simultaneously twisting its head from below and presenting the black throat patch—typical for this species—to its mate. This is an important releaser for mutual preening (see Section 8.2.5). In many Palearctic and African finches the male sits next to the female after successful copulation and assumes the female copulatory posture, where the female in some species may even show intentions of mounting. Mallard drakes also show a distinct female courtship movement after copulation, the so-called nod-swimming, with which they circle the female. In the mute swan, both birds rise out of the water, their breasts together, while at the same time waving their downward-pointed bills two or three times from side to side. The biological significance of postcopulatory displays seems to be the mitigation of increasing aggressive or escape tendencies that occur after copulation, thus making continued association of the partners possible. Hence, these behavior patterns strengthen the pair bond at the same time.

According to another interpretation, postcopulatory displays are said to discharge excess energy. This view needs further supporting evidence. It does not seem very probable, since it is hard to imagine a selection pressure for such complex behavior sequences merely for the release of excess energy.

8.2.2.4. Synchronization

In contrast to humans, in many domesticated animals, some tropical vertebrates, and some animal parasites, reproduction is limited to a certain time of the year. There are two reasons for this temporal restriction: (1) The production of germ cells and the various behavior patterns involved in caring for young make special physiological demands on the adult organism. (2) If the young are to have the best chance of survival, they must be born at a time when the food supply is qualitatively and quantitatively optimal and readily accessible.

The reproductive period of each individual species or population has been adjusted by natural selection during the course of evolution to those months of the year that best meet these preconditions. Furthermore, additional factors beyond food supply may have had an additional influence, e.g., temperature, moisture content in the air and soil, opportunities for egg laying and care of young, as well as other environmental factors. The time of readiness for reproduction can be very long and can last for several months, as in many tropical species, or it can be limited to a few days or even hours per year, as in the South Pacific.

Paralleling the temporal limitation of reproduction, as expressed by physiological processes, is the expression of sexual behavior, which as a rule occurs only at certain times of the year. Sex hormones have a controlling influence in this respect.

An important biological problem is the temporal synchronization between the organism and periodically occurring events in its environment. This is especially obvious with respect to reproduction: Birth or weaning of young is the end of a developmental sequence that began with fertilization of the egg and continued with subsequent embryonal development, nest building, and other forms of parental care that, depending on the size of the animals, can take many weeks or months. In many instances these processes must begin well before the onset of suitable conditions and must run their course. This in turn presupposes that the organism be informed early about the impending improvement of environmental conditions. The environmental stimuli that trigger the reproductive cycle sooner or later before the birth of the young, and that at the same time program the necessary physiological growth processes, are as a rule not linked primarily to the availability of food or other factors needed for the successful raising of young. Their place is taken by other "forewarning" factors that announce the arrival of conditions suitable for reproduction. These include the annual changes in day length in geographic locations that affect a great many species. Many bird species are especially sensitive to an increase in day length, which affects their gonadal growth and the recurrence of sexual behavior. In the biological literature, the terms PROXIMATE (or immediate) and ultimate have led to the evolution of specific reproductive periods.[6]

Next to day length and a number of other environmental conditions, sexual behavior itself can constitute an important immediate factor. This is biologically meaningful for the following reason: Through other environmental factors, especially when these exhibit marked changes during the course of the year, reproductive cycles within a population are largely synchronized. However, certain individual differences remain that may be due to age differences or varied experience. Successful mating, and especially the external fertilization in many freshwater and marine animals, where the sperm and

[6] At first, these terms were used only to refer to annual periodicity. Today they are also applied to other examples in which there is a divergence between phylogenetic selecting and actual regulating factors.

eggs are ejected into the water, require a temporal fine-tuning of both partners' activities. Such coordination cannot be assured by the usual environmental factors but can be achieved only through the mutual influences of the animals upon one another. Thus, we see here an additional function of sexual behavior in the synchronization of both sexes. In promiscuous species, i.e., in which male and female form no permanent pairs but meet only for copulation, this synchronization is brought about by various forms of precopulatory behavior. In animals that form long-term bonds, this can be accomplished by pair-bonding behavior patterns that are often much less "conspicuous" (see Section 8.2.3).

Temporal synchronization as a rule is achieved by the stronger reaction of one sex to environmental stimuli, which in turn stimulates the development of sexual behavior in the other. As a rule, the male is more responsive to environmental effects. Investigations on European and North American song-birds have shown that in the male an increase in the daily light phase, i.e., in artificial day length, will lead to gonadal growth, increased singing, and the release of sexual behavior patterns. In the female, on the other hand, the same conditions lead only to an incomplete development of sexual behavior, while the complete development requires the influence of additional factors, among which behavior plays an important role. This is in agreement with the observation that the annual reproductive phase is frequently longer in the male than in the female. In many bird species the gonads of the males develop earlier in the spring than do the ovaries of the females, and in migratory species the males arrive a few days earlier in their breeding grounds and with more developed gonads. A large number of experimental data support the hypothesis that male courtship behavior stimulates females. In some mammals, e.g., shrews, rabbits, European ground squirrels, nutria, and European polecats, the female ovulates only after successful copulation. In the rabbit this is about 10 hours after mating. This so-called induced ovulation insures that ova ready for fertilization are available only when sperm is present. Here, copulation itself is responsible for the temporal synchronization. Other sexual behavior patterns can have a similar effect. It was found that female ringdoves kept in isolation ignored a nest in their cages and laid no eggs. If a male was added, eggs were laid within a few days. The same effect is a achieved by presenting a male that is kept separated from the female by a glass partition. However, a castrated male that showed no courtship behavior did not induce egg laying. In budgerigars, auditory stimuli play an important role: Females that were isolated from males but could hear their vocalizations began to lay eggs within 18 days, while females without auditory contact with males had not laid eggs even after 30 days from the start of the experiment. Similar results were obtained for other dove species and for some songbirds (starlings, English sparrows, and canaries). In many mammals the influence is olfactory stimulation: In mice a male is able to influence ovarian development in a female positively if she is placed in his cage inside a closed cage of her own. Even exposure to a cage in which a male has lived can have the same effect.

In some instances, especially in the ringdove, the role of hormones is

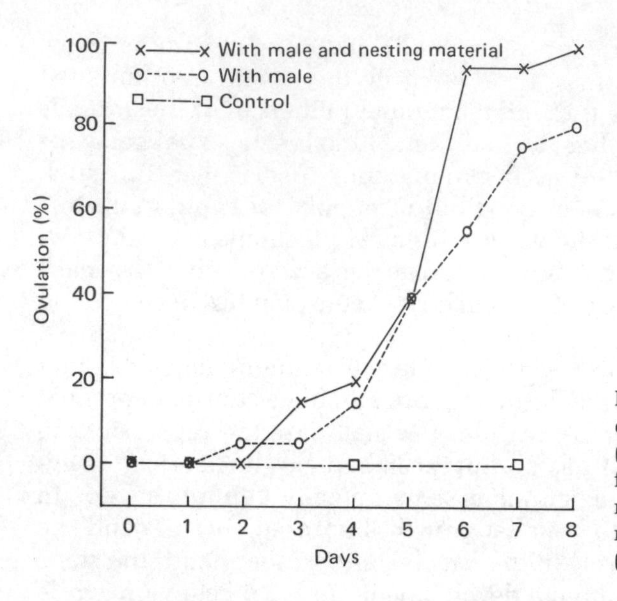

Fig. 69. The influence of the male and of nesting material on the ovulation (bursting of follicle) in ringdoves. The females of the control group were maintained without a male and nesting material in otherwise identical cages (after Lehrman 1964).

already known. The secretion of hormones in a female is triggered by a courting male. This in turn leads to ovarian development, egg laying, and nest-building behavior (see Section 5.2.2.3). It can also be assumed that synchronization of the sexes takes place in other species, especially in those where the partners engage in prolonged courtship behavior sequences in which both animals show either very similar or identical behavior. Examples are the courtship flights of some hummingbird species and species in which an action of one sex releases a subsequent behavior in the other sex, such as in the courtship behavior sequence of the three-spined stickleback.

This stimulating and synchronizing effect of courtship behavior is not always limited to the partners of a pair. In some instances it can also affect other pairs and influence their readiness to mate via the usual interaction between hormones and behavior (see Section 5.2.4). Hence, entire groups of conspecifics become synchronized in their courtship activities.

There are numerous observations and experimental support for this. In many colony-nesting birds, the pairs in a colony or general area begin to lay eggs and incubate at about the same time, whereas between colonies and areas there are larger differences. Egg laying in large colonies often begins sooner than in small ones and earlier in the densely settled center than at the periphery.

These differences usually cannot be accounted for by the usual differences at the breeding site but can be traced to the mutual influences of the birds on each other. It is known in the budgerigar that the gonadal development of male and female animals can be facilitated by the presence of courting neighbors. In ringdoves the auditory presentation of colony sounds to females in isolation cages stimulated ovarian development.

In some species the synchronizing effect is further increased by the fact that courtship behavior is not limited only to members of one pair, but that many animals join in a communal courtship. A well-known example for this is the "ceremony" of flamingos, which consists of stretching and preening behavior patterns and which involves dozens or even hundreds of birds.[7]

While the biological significance of mutual synchronization of male and female primarily insures fertilization, it is less uniform in the temporal coordination of activities beyond the individual pair. In some water animals the same purpose may be met, since the chances of fertilization are increased when many males and females release their sex products into the water at the same time. However, in colony-nesting birds, other reasons could play a role. Here, close synchronization of egg laying offers better protection against predators in two ways: (1) The simultaneous presence of all colony members can increase the effectiveness of joint attacks on the predator. (2) There is an overabundance of potential food for the predator during the breeding season in the colony, but the actual time when eggs or young are present is shorter; hence, the total loss to predation in the colony is numerically less than if beginning and end of reproduction were separated by a longer time span. Temporal synchronization can further insure that all pairs are in the same

[7] As in the case of simultaneous courting by several males so as to attract females (see Section 8.2.2.1), the frequently used term *social courtship* should also be avoided here.

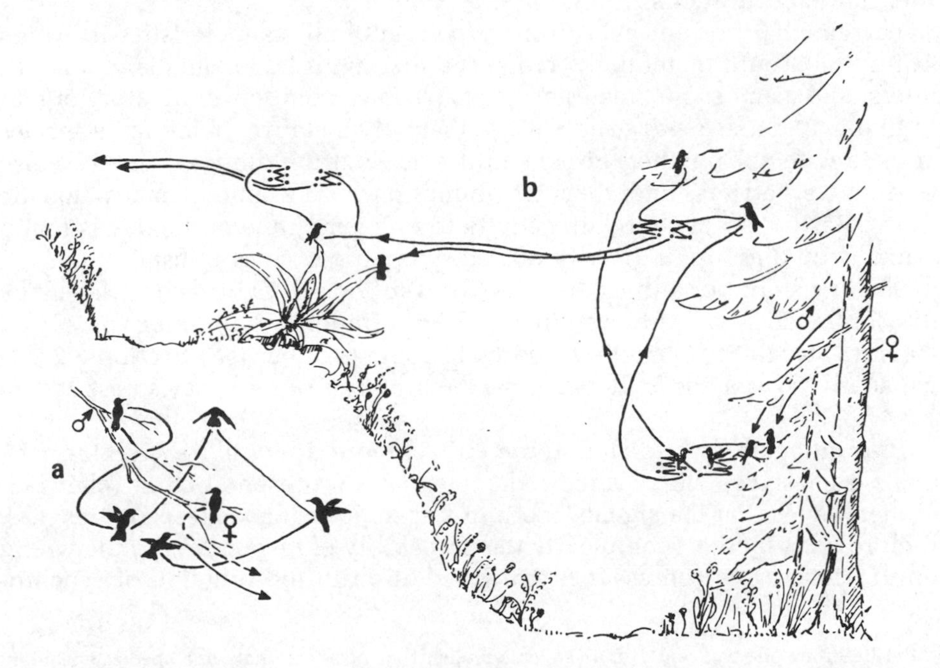

Fig. 70. Courtship flight of the hummingbird (*Hylocharis leucotis*). During the first phase, the male circles the female in an undulating flight (a); in the second phase, both partners fly around together, often hovering in front of each other (b) (after Wagner 1954).

stage of a breeding cycle (nest building, incubation, raising young), and the total number of mutual interruptions between neighbors is less than would be the case of one pair was still building while another was already incubating. In climates in which the seasons are irregular and obvious environmental stimuli for gonadal development are absent (see the beginning of this chapter), as in many parts of the tropics, the additional value of social stimulation can readily be seen. Here pairs can affect their breeding condition by increasing mutual courtship, and hence achieve a rapid onset of reproduction as soon as environmental conditions are favorable. In similar fashion the communal courtship of flamingos probably achieves a simultaneous level of reproductive readiness so that the birds then gather and found a breeding colony.

8.2.2.5. Sexual Isolation

Mating between different species is biologically disadvantageous. As a rule, this results in offspring that either are not viable or are not able to reproduce; hence, it is a "waste" of the reproductive potential of a species. Even in cases of usually closely related species that can produce fertile hybrids, a mixing of genomes frequently leads to disruptions in the harmony of the behavior and environmental adaptations, because the gene complexes are not balanced (see Chapter 10). Hence, there is a strong selection pressure against the crossing of species, and hybridization is rare among most species under natural conditions.

Barriers that prevent hybridization consist of all characteristics in which species that could potentially cross are distinguishable. In many insects, spiders, and many snails, males and females have complex copulatory organs that fit together like a "lock and key" so that only members of the same species can mate with one another. Seasonal differences in reproductive periods also are effective barriers, and they are found especially among many aquatic species. Restriction of courtship activity to a specific time of the day is found in small fruit flies and in closely related sympatric species of fish.

The most important role in the prevention of hybrids is played by behavioral barriers. Here we find a fourth important function of sexual behavior. Signals that enable the sexes to come together (see Section 8.2.2.1) are usually so specific that only members of the same species respond to them.[8]

Prevention of hybridization can be achieved in two ways: Either the signals themselves differ and/or the manner of presentation is changed. Another way is that the signals occur in a specific sequence (see Section 4.1), which reduces by their complexity the probability of responding to the wrong stimuli. Hence, one can assume that in addition to the function of synchro-

[8] In the literature one frequently finds statements that the specific signals of a species enable the animal "to recognize its species." Ernst Mayr has correctly pointed out, however, that this concept presupposes awareness, which the great majority of animals almost certainly do not have. Hence, this formulation should be avoided.

nizing the activities of the sexual partners (see Section 8.2.2.4) by complex courtship sequences; isolation of species is a factor as well.

The danger of hybridization is greater the more closely related two species are to one another, i.e., the more closely they are arranged in the zoological classification system. Special problems of species isolation exist in situations where several related species live sympatrically, i.e., in the same area. Here the sexual releasers need to be unique so as to avoid mistakes (see Section 3.4). Thus, the courtship plumage of male ducks in North America and Europe, where many species live together in the same habitat, is very conspicuous and varied. In regions in which only a few duck species live, and where the possibility for a "mix-up" is less, the males are often cryptically marked, i.e., they carry few if any distinct releasers. Such species differences in signals can also occur within one species: On some Hawaiian Islands the males of the mallard duck, which on the mainland are conspicuously colored, exhibit a cryptic femalelike plumage the year around. Some bird species, living in areas with several close relatives, are characterized by well-differentiated songs, whereas in areas with fewer sympatric species, such as on small islands, their songs are much more variable. Examples are the blue titmouse on Tenerife and the goldcrest in the Azores. Some of the most impressive examples of precision and specificity of signals in sympatric species are the flashes of lightning bugs (see Section 8.2.2.1); the sounds produced by many grasshoppers, frogs, and toads; or the claw movements of fiddler crabs, which can be distinguished by their distinct forms and rhythms (see Section 8.2.2.1).

More recent investigations have shown that a certain sexual isolation can even exist between various populations of the same species. This might prevent the loss or dilution of some subtle, population-specific adaptations to specific environmental conditions (see below), whether these be genetically determined, based on learning processes (e.g., habitat imprinting, see Section 7.4.7.2), or passed on by tradition (see Section 7.5), and which would be lost again if members of different populations produced offspring together.

Even in intraspecific isolation, behavioral factors can play an important

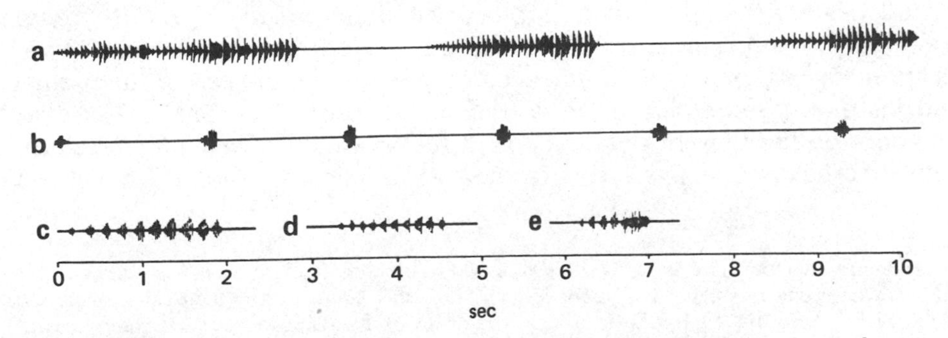

Fig. 71. Songs of five species of grasshoppers. The two species—*Chorthippus biguttulus* (a) and *Ch. brunneus* (b)—are closely related and live in the same habitat. Their songs are greatly differentiated. The species *Ch. montanus* (c), *Ch. longicornis* (d), and *Ch. dorsatus* (e) are also closely related but live in very different habitats. Their songs are quite similar (after Jacobs 1966).

role. In various species of birds, song dialects have been more closely investigated. A DIALECT is a specific form of vocalization that is distinguishable from the vocalizations of the same species in another area.[9] Although the exact function of dialects is not yet clear, it can be assumed—at least in some instances—that they help to preserve the special adaptations within a species. This is true especially for species in which several "ecotypes" have evolved, i.e., in which in some populations differences can be seen between food and biotope preferences that are more or less well developed (see footnote to Section 8.2.1). Here a dialect can lead to the "correct" mating, i.e., it can facilitate the meeting of partners of the same adaptive type, when females preferentially mate with males whose song type corresponds to that of their father or his territorial neighbors. Furthermore, a dialect may be an important cue in finding the home range of a particular population when the birds return from their winter migration. Clear indications of this are found in the ortolan bunting, where several ecotypes each have their own specific dialect. Observations from other species that also developed dialects also point to the isolating effect of such a dialect mosaic.

8.2.3. Pair-Bonding Behavior

In many species the sexes meet only for copulation; in others they remain together for some time and form pairs. Long-term bonds are known in all vertebrate groups, and even in some crabs and insects.

The biological advantages of the sexual partners remaining together are many. First, such an arrangement is needed in all cases where both sexes participate in raising their young (see Section 8.3.2). It can also have an additional important function in the reproductive process: it can make new pair formations unnecessary and hence behavior patterns that facilitate bringing the sexes together (see Section 8.2.2.1). It can help the synchronization of the partners, and, finally, it can make it harder for species to cross and thus favor sexual isolation.

Where male and female already live together before the reproductive period begins, the annual development of their gonads is more strongly coordinated by the similarity of the controlling environmental factors, as well as by the mutual behavioral influences, than would be the case if the partners had just met (see Section 8.2.2.4). Such synchronization may have a direct influence on the time of reproduction. Thus, in the kittiwake gull, pairs that have bred before lay eggs earlier in the season than pairs that have just been

[9] This definition shows that the term *dialect* is used in ethology and bioacoustics more generally than in everyday language, which for humans includes languages as well as dialects. Most recently the term *dialect* has been used in the behavioral sciences to include cases in which intraspecific differences are found within a communication mode. Hence, one speaks of chemical dialects in the sex attractants of female butterflies (see Section 8.2.2.1) and of gesture dialects in macaques, where in various troops the same information is transmitted by different, troop-specific head movements.

newly formed. Females that have changed their mate since raising their last brood are less successful the next time. They lay fewer eggs, and a smaller percentage of them develop into fledgling young.

The long-term synchronization within a pair at the same time assumes one of the important functions of courtship (see Section 8.2.2.4). Accordingly, as was already observed in doves, courtship behavior, i.e., precopulatory behavior, is comparatively shorter in established pairs; it is less conspicuous and gives the impression of "efficiency" compared with newly mated pairs.

The possible connection between pair formation and sexual isolation can be seen especially in those groups that include species with and without long-term bonds between the sexes. In the birds of paradise in New Guinea, where hybrids are relatively frequent, they have been found and described for a larger number of species and genera. However, they are found only in species in which the female, when ready to mate, seeks out the male on his courtship arena. After mating, she builds her nest alone, incubates the eggs, and raises the young. Of the three species in which male and female form pairs and where the male also helps to raise young, no hybrids are known. Such a high frequency of hybrids is also known in other groups of birds that have no pair bonds between the sexes, especially in hummingbirds, manakins, and grouse. We can, therefore, be certain that long-term bonds between sexes contribute importantly to the prevention of matings between species.

8.2.4. Forms of Heterosexual Bonds

As a rule, a heterosexual bond is only found within a pair, i.e., between a male and a female. Such MONOGAMOUS bonds, which may last sometimes for years or even for life, are found in all classes of vertebrates and sometimes in invertebrates, e.g., in some isopods and in some species of crabs. In birds, monogamy is especially widespread, but it is rare in mammals and has been conclusively demonstrated only for some members of the dog family, some rodents, and some hoofed animals, as well as in several primates, e.g., gibbons, titi monkeys, and marmosets.

In some instances a permanent bond can extend beyond a singular pair and include several members of another sex. If one male is paired with several females, one speaks of a HAREM or ONE-MALE GROUP. Harems have been described especially in mammals, e.g., in many primates, in steppe zebra and mountain zebra, and among fishes in African and South American cichlids.[10] The opposite relationship, for which there is no separate name, is found in

[10] Besides the given examples, there are some mammals, especially in many pinnipeds, where temporary liaisons occur between one male and several females that are frequently called harems. However, the liaison is very short in these species and is only for the purpose of mating. Furthermore, the composition of the group may change continuously, or the entire group may be taken over by another male. Since a permanent bond between individuals seems not to exist, the term *harem* should not be used in this case.

the Tasmanian water hen. Here as a rule, two males are paired usually with only one female.[11]

8.2.5. Mechanisms of Maintaining Bonds

The long-term bonding between members of a pair can be basically accomplished in two ways: (1) through relationships with the environment and (2) by bonding of the partners to each other. An impressive example of the former is given by some crabs that are forced to live together for life: In this species young males and females enter a sponge or coral cavity together and spend the rest of their lives there. Gradually they grow to such a size that they are unable to leave the cavity; hence, special bonding mechanisms are not necessary. In many migratory birds, in which male and female return each spring from their wintering grounds, where they live independently of one another, their bond is mediated by each individual's affinity for the common breeding place. This form of pair bonding, which is somewhat similar to an aggregation (see Section 8.4.4.1), is called *Ortsehe* (translated "wedded to a place") or—since the partners apparently do not know one another—"anonymous monogamy." In most instances of permanent pair maintenance there seems to exist a genuine bond between its individual members. This presupposes individual recognition (see Section 8.4.4.2), where the male behaves differently toward his partner from the way he behaves toward other females. In this case, certain behavior patterns can contribute to the maintenance of their bond. Their form and origin vary with species. Because of the wide distribution of monogamy in birds, it has been extensively investigated in this class of vertebrates. Here two mutually exclusive regularities can be identified: (1) Frequently pair-bonding behavior is part of courtship behavior or (2) frequently there are behavior patterns that are performed together by the pair in alternation or at the same time.

Sometimes behavior sequences, which occur in "true" courtship, e.g., leading to copulation, can also maintain a bond. Australian zebra finches frequently court one another outside of the reproductive period as well. The behavior patterns are the same as those preceding copulation, but they do not result in actual copulation. In many primates, e.g., baboons and rhesus macaques, copulation can reinforce the bonds even between individuals of different sexes. Since in these species a series of copulations are needed until

[11] Sexual relationships between one individual of one sex and several individuals of another sex are known in many species. These are called POLYGAMY, and one distinguishes between POLYGYNY (one male having many females) and POLYANDRY (one female having many males). Within the class of birds, both conditions occur in addition to monogamy and promiscuity (see Section 8.2.2.4). However, there is almost exclusively successive polygamy involving only relatively short bonding periods between the sexes. For example, a male or female may successively mate with several males and females, with the last-named sex usually caring for the young alone. The Tasmanian water hen provides the only example within the birds in which there is a permanent and simultaneous bond between more than two sexually mature animals, which can be compared to the harems of mammals—although the sexes are reversed.

an actual ejaculation occurs, and since these are separated by intervals, the single copulation can be considered a bonding mechanism. The same observation has been made in the Indian flying fox. There three to seven copulations are needed before an ejaculation occurs. These sequences are seen only during the reproductive period; single copulations, however, are seen throughout the year, even involving pregnant females.

In other instances individual "emancipated" partial actions of courtship can occur. They seem to serve the bonding of the pair, and they also occur outside the reproductive period. This is true for courtship feeding (see Section 8.2.2.2), which is found in many parrots and in the raven throughout the year.

Duets and antiphonal songs performed by both sexes are probably among those behavior patterns that bond partners of a pair together. They are especially widespread in tropical bird species, but they also occur occasionally in monogamous mammals, e.g., gibbons. In these species either the partners of a pair sing simultaneously or they alternate. In the more primitive examples, the vocalizations of both sexes are probably identical. In the more specialized and advanced species, male and female produces differentiated stanzas, or segments thereof, which are combined in some specific manner. Some African shrikes produce some very highly developed songs. They are characterized by a surprisingly short reaction time of the joining partner; e.g., in the gonolek, which holds the record in this respect, the end of first partner's stanza is followed within 125 milliseconds by the onset of the second partner's. Since these duet songs are also produced outside the breeding season, one suspects that they serve to maintain the permanent pair bond between the partners. In addition, they probably aid in locating the partner, especially in species that are active at night or who live in dense vegetation.

In mammals and birds the behavior involving the mutual grooming or preening of fur or feathers is found among members of a pair. Birds use their

Fig. 72. Preening white-eyes.

beaks, primates use their hands, and many other mammals use their tongues or teeth. Since social grooming is primarily directed toward those parts of the body that an animal has difficulty in reaching, it can be assumed that the function of the behavior is to clean and to remove parasites. In many preening bird species no actual cleaning has ever been observed. It is possible that the original function of preening or grooming was to clean, and that during the course of evolution a change of function to facilitate pair bonding became increasingly important. Contact behavior, i.e., active touching of bodies, which in many species occurs only or preferentially between members of a pair (see Section 8.1.6), may also help to maintain bonds. The same purpose can be served by nest-building behavior and other behavior patterns associated with the nest if they also occur outside the actual nest-building season. For example, in some Australian waxbills the male indicates to the female the potential nest site by specific movements and calls. In some species, so-called nesting ceremonies have evolved that consist, in part, of completely synchronous movements of both partners, which are carried out not only prior to nest building but also throughout the breeding cycle and even outside the breeding season.

Since many of the behavior patterns that facilitate pair bonding occur with increased frequency when the partners have been temporarily separated e.g., when searching for food, this behavior is often referred to as GREETING BEHAVIOR.

8.2.6. Motivational Aspects

If a pair is maintained by courtship behavior patterns, then it could be possible that the behavior is sexually motivated. When pair-bonding behavior occurs outside the reproductive season as well, hence at a time of gonadal inactivity (see Section 5.2.2.1), or if—as in the case of nest-building behavior—they are not closely connected with courtship behavior, then a purely sexual motivation is hard to imagine. For this reason the hypothesis has been expressed that pair-bonding behavior may have its own independent motivation; it is then called a BONDING DRIVE. Actual evidence for this is known only in two species to date: the "triumph ceremony" of the greylag goose and the "spatial bond" (sitting together) in a tropical shrimp. In both instances, the behavior is clearly different in form from all other behavior patterns, and it is temporally independent from all aggressive, sexual, or care-of-young behavior. In other words, the behavior cannot be classified among the three classical drives of aggression: escape, sexuality, or care of young. As soon as the partners are separated from one another, they look for one another, so that one may postulate a specific appetitive behavior. Since the greylag goose and the shrimp are two entirely different animal groups that have a bonding drive, it is conceivable that there may be still other species with an independent motivation for pair bonding in instances where partners remain together for a long time.

8.3. Behavior in Care of Young

8.3.1. Functions in Care of Young

In many species the prerequisites and conditions for development of young are actively created or improved in various ways by the parents. This occurs either by directly caring for the young—i.e., by behavior patterns that protect, feed, or care for eggs and young—or by a kind of preparation for care—i.e., by building protective structures, nests, or cocoons, by the gathering of a one-time food supply, or by the deposition of eggs in protected areas or near suitable food sources. The latter form of care, in which parents and young do not come into contact, is usually called provisioning (in German, *Brutfürsorge*), or care of eggs and young in cases where direct contact is involved. However, these distinctions are not always made. The common biological goal of provisioning and care of young is to insure or increase the chances of survival of one's own young, and hence the contribution of one's own genes in the next generation (see Section 8.4.3.2).

Warning of danger, hiding and guarding of eggs or young, and their active defense all protect the descendants. A very conspicuous behavior is the distraction display called "broken-wing ruse," shown at the approach of predators by many ground-nesting birds. Here, the parent bird tries to "mislead" the predator by limping ahead—wings dragging, seemingly broken—hence promising an easy meal for the predator. When the predator has followed the bird far enough from the nest, the bird flies off, eventually to return to it or the young.

Many predators and rodents carry their young to a safe place when they sense danger by picking them up carefully in their mouth or with their teeth. They also retrieve them when they find them outside the nest or den. This retrieving is of general interest since it is one of the few examples of a behavior that does not become fatigued (see Section 2.3). Carrying of young occurs also in other animal groups. It is either done in special pouches, as in most marsupials, or by the young sitting on or clinging to the parents, as in many spiders, scorpions, the crested grebe, and many mammals, e.g., the koala, anteaters, sloths, bats, and primates. Many young mammals actively cling to their mother or other group members with their hands and feet where so-called grasping reflexes can be discovered when hands and feet are touched. Bats and some primates also hold onto their mothers with their lips on her nipples.

An impressive form of body contact between parent and young is that found in mouthbreeding fishes: These animals store and carry the eggs in an expandable mouth, where the young are continuously supplied with fresh, oxygen-rich water as the parent opens and closes its mouth. The young remain in the mouth for some time even after hatching, or they return there from their "exploratory forays" whenever there is a sign of danger. Mouthbreeders are found especially in cichlid fishes, but also in some other groups, e.g., catfish, anabantids, and killifishes.

Feeding of young is accomplished either by giving them food directly or by the parents leading the young to suitable locations, or perhaps actually showing them what to eat. Such "food calling" is especially found in some precocial birds (see Section 7.4.7.1). Direct presenting of food may consist of the parent merely bringing the food or actually feeding the young. Food may also be produced within the parent's body for the young. This may be the milk of mammals, crop-milk of pigeons and doves, stomach fluids in flamingos, mucous skin secretions in discus fish, and gland secretions of worker bees.

In warm-blooded animals such as birds and mammals, the transmission of body heat from the parent to the young or the eggs is an important function of parental care. This behavior is performed until the young animal is able to thermoregulate on its own.

In birds, keeping the young warm is called BROODING. In many fishes the supply of oxygen to the eggs is improved by the "fanning" movements of the parent(s), which uses its fins for this purpose. Parental care includes the cleaning of young and removal of parasites or fecal material from the nest in many bird species.

These direct actions on behalf of the young are extended by the transmission of information from parents to offspring. An important example is the song of birds. In most species this must be learned, and this learning process takes place quite early in life. Frequently it occurs before the young are independent, and in some species while they are still in the nest. It was shown in bullfinches and zebra finches that the father's song is the model for the young male. In this way the specific form of a song is passed from one generation to the next. The knowledge of food, the techniques of acquiring it, the recognition of predators and other dangers can be transmitted by the parents in various ways. In higher mammals, direct imitation of adults plays an important role in this respect (see Section 7.4.5).

Overall parental care is an important means by which traditions are developed through the direct contact between two generations (see Section 7.5). This task of parental care was probably acquired secondarily during the course of phylogeny. However, it has become so important in some mammals today that the young remain for some time with their parents long beyond the

Fig. 73. Leopard retrieving young.

Fig. 74. Hamadryas baboon mother with young. The young clings to the mother and holds on at the nipples with its lips.

time of food dependence. In the three-toed sloth, the young is nursed only for one month but is carried on her back for another five by the mother. During this time it learns about food plants, resting places, and other environmental contingencies. A longer association of parents and young, and young with each other, has also been described in some other mammals and in some bird species that live a long time. There have been descriptions of "helpers" at the nest in some birds (see Section 8.4.3.1). There are indications that the acquisition of information is a factor here as well.

8.3.2. Formation of Families

Care of young is found in almost all animal groups. Even in the most primitively organized multicellular animals, the sponges and colenterates, simple forms of care of young are known, but among the lower invertebrates it is rare. However, in two groups—arthropods (e.g., crustaceans, spiders, millipedes, and insects) and vertebrates, parental care is widely distributed. In the highly developed groups, the termites among insects, the birds and mammals among vertebrates, it is present in all species.

The participation of the adults in caring for young varies, however. The mother or father may each alone care for their offspring, or both parents together, or an entire group of adults may provide care. Hence, the term FAMILY is used purely descriptively in ethology and refers simply to the living together of parents and young. Depending on the relative contributions of parents in caring for their young, one distinguishes among parental, female, and male families; additional subdivisions in these categories may be made.

The parental family, in which both sexes together or in alternation care in the same manner for their young, exists in the majority of bird species, but it is also found in fishes, especially in many cichlids, and in some mammals such as gibbons, marmosets, and canine predators. If there is a division of labor between the two parents, one speaks of a MALE–FEMALE FAMILY. It is typical for many raptors and hormbills, in which the male brings the food while the female distributes it to the young. If the division of labor is such that only the female cares for the young while the male "merely" defends the territory or guards the family, as with many songbirds and mammals, then one speaks of a MALE–MOTHER FAMILY. There are all kinds of transitions to the pure mother family.

The MOTHER FAMILY is the typical family structure of most mammals in which the mother, because she produces milk, has the inside track, so to speak, in caring for the young. This arrangement is also found in birds (e.g., in many fowl, hummingbirds, and birds of paradise), fishes, spiders, and scorpions. The father family, in which only the male cares for the young, is only rarely found, as in sticklebacks, anabantids, and sea horses, and, among birds, in phalaropes, button quails, emus, and rheas.

The insect societies are also organized into families: among the sawflies and wasps (wasps, bumblebees, bees, and ants), into mother families; in the termites (where the male remains with the queen even after the nuptial flight), into parental families. In some species other conspecifics help to care for the young in addition to the parents. These GROUP FAMILIES will be discussed elsewhere (see Section 8.4.4).

8.3.3. Behavioral Adaptations with Respect to Care of Young

As in other areas of social behavior, care of young requires means of communication between conspecifics. To meet this end, various and sometimes very complicated behavior patterns have evolved. They serve primarily two functions: the reduction of intraspecific aggression, and the initiation and completion of the presenting of food.

Aggressive motivation can occur between the parents and between parents and young. Within the parent pair there exists a special danger in all those aspects where care of young entails a physical approach or actual contact between the partners. Some species have special adaptations in the form of behavior patterns that can be recognized as aggression-inhibiting behavior and that have a function similar to courtship and pair-bonding behavior. First among these are the "relieving ceremonies" in many bird species that occur when members of a pair exchange places at the nest with an incubating or brooding partner. Some species carry nest material in their beaks when relieving each other. The aggression-inhibiting effect of this behavior, which shows similarities to "nesting symbols" during courtship (see Section 8.2.2.2), is especially obvious in the flightless cormorant of the Galapagos Islands. When returning to the nest, the male carries a sea star or a bunch of seaweed, which is accepted in a clearly aggressive manner by his mate. If one takes such material away from the returning bird, then the mate will direct her aggression directly to her returning partner, who may actually flee because of the severity of the attack.

The young can also bring out aggressive tendencies in their parents. These can be suppressed in two ways: The young can show behavior patterns that appease, or they can avoid the sending of aggressive signals as much as possible. Juvenile begging behavior may have such an appeasing function in addition to releasing the actual feeding behavior. Hence, we suspect that because of this function, begging behavior may also have become incorporated into other areas of social behavior in which aggressive tendencies need to be reduced (see Sections 8.2.2.2 and 8.4.6).

Avoiding the presentation of aggressive signals can be achieved by special "juvenile" plumage patterns. These have a dual function: On the one hand, they are very inconspicuous and offer increased protection through effective camouflage. On the other hand, they are characterized by a complete lack of signals that in adults play a role in aggressive displays, e.g., in threats; hence, they release little or no aggression.

Beyond the functional system of family and care of young, there is a special function inherent in juvenile characteristics, especially in group-living species, in which the young occupy a status within the group. They are afforded certain special considerations, e.g., as in squirrel monkeys where they are not yet integrated into the social rank order (see Section 8.4.6.1). This also affords them the opportunity for much learning, which becomes possible when all group members recognize the young as such.

There is experimental verification for this hypothesis in the mute swan. In this species the young usually have a gray-brown plumage, but a small percentage of young is white like the adults. In experiments involving brown and white models, the latter were threatened and attacked more frequently by territorial swan pairs than were the brown models. This supports observations in the wild in which aggression, even by their own parents, was greater toward their white young than against those with the more usual brown juvenile plumage.

The reverse can also occur in that the parents trigger aggressive tendencies in their young. In this instance, they have aggression-inhibiting behavior patterns. This is found in the night heron: When a parent lands on the edge of the nest to bring food to the young, it bows deeply and erects two or three usually depressed head feathers, which offer strong contrast to the blue-black color of the top of the head, thus presenting a very conspicuous signal. Only then does the parent enter the nest. If one prevents the parent from bowing, he or she will be attacked by the young.

Beyond the area of intraspecific aggression, there is danger in predatory animals that they confuse their own young with prey. Here, too, special adaptations evolved. For this reason some spider species do not eat when they care for their eggs and young, and a similar inhibition is seen in some mouthbreeding fishes.

The passing of food from parents to young requires a high degree of coordination and temporal synchronization between the partners, which, in its basic elements, is parallel to the initiation of copulatory behavior (see Section 8.2.2.4). In both instances, several behavior sequences and releasers have evolved. Again they are most evident in birds where feeding involves direct contact from the beak of one bird to that of the other. Such releasers are the gape markings in the throats of young, which are often brightly colored, the begging calls, and the sum total of behavior patterns of young that are commonly referred to as begging behavior. Their task is the release and directing of parental feeding patterns.[12] On the other hand, parents send out behavioral and morphological signals that release begging behavior in their young. They have been experimentally investigated quite early in the history of experimental ethology, because they are especially well suited for the analysis of innate releasing mechanisms. The best-known example is probably the explanation of the function of the begging-releasing characteristics of the herring gull beak (see Section 1.3).

8.4. Behavior of Groups

In many species adult, sexually mature species members form temporary or permanent groups beyond the formation of a pair. The function, age, and sex composition of such social groups can vary from one species to the next. In spite of this, some general regularities can be identified.

8.4.1. Biological Function of Social Groups

The fact that the formation of groups occurs in many species of varied systematic classification indicates that they have an important and probably varied biological significance. The advantages offered by group living, which

[12] It must be pointed out, however, that the exact function of begging behavior in animals has not been thoroughly investigated in a quantitative or experimental fashion.

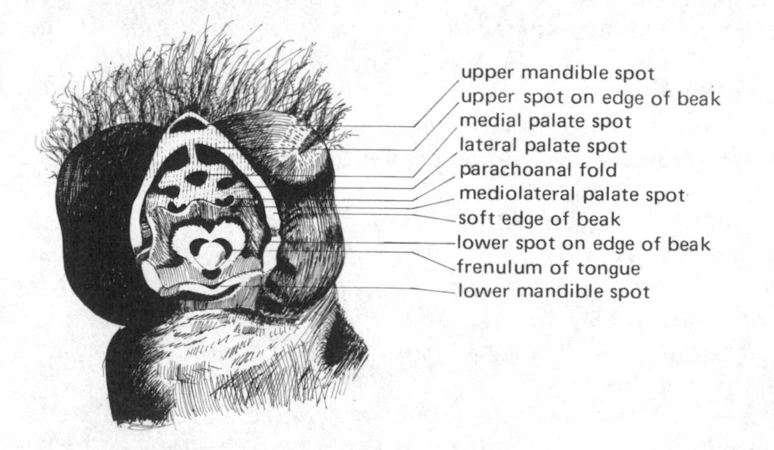

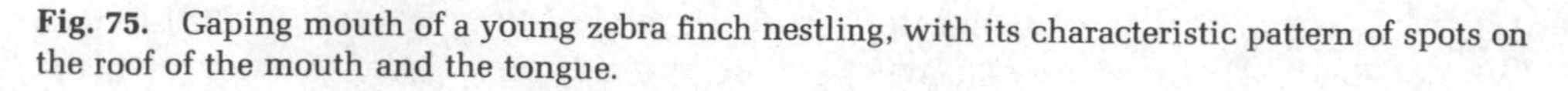

Fig. 75. Gaping mouth of a young zebra finch nestling, with its characteristic pattern of spots on the roof of the mouth and the tongue.

Fig. 76. Night heron with normal body posture (left) and erected head feathers after landing on the edge of its nest (after Lorenz 1935 from Tinbergen 1955).

offer a possible selection pressure that may have facilitated the evolution of social structures, are many.

8.4.1.1. Avoidance of Predators

Living in groups offers the individual animal a better protection against predators in three ways:

a. Mutual Warning. The most widespread benefit lies in the possibility of mutual warning. A group is always as alert as its "best" member, i.e., the one that reacts the fastest. While in the protection of the group, an individual animal can afford to be not alert for a moment and to engage in some other activity. Mutual warning can involve several sense modalities. Acoustic and chemical signals are especially frequent. Examples are the warning calls of birds and the alarm substances of fishes. Sometimes warning signals can be effective beyond the species, a fact that underlines the importance of mutual warning. A well-known example is the similarity of the alarm calls of various songbird species (see Section 3.5).

b. Confusion Effect. Many species, e.g., birds, fishes, bunch up as soon as a predator appears. The apparently increased difficulties in focusing on one individual by the predator, and the fact that some that are not very hungry may be deterred from attacking, increases the probability of survival for each individual member of the group. Predators in turn have adapted to this behavior of their prey species by trying to separate individuals from the group. In experiments, sticklebacks preferentially caught daphnia that were farthest away from the swarm. In the presence of closed groups of prey, typical conflict behavior was observed in both stickleback and pike.

c. Common Defense. In large numbers even "helpless" animals can put up a good defense. Hence, we can observe in many species an active closing of ranks for the purpose of defense. Musk-oxen, wild sheep, and bison form defensive circles in which the horns of the adults are directed outward while the young are protected inside the circle. In baboons the adult male animals interpose themselves between their own group and the predator. They are able to frighten away even jackals, dogs, and leopards by joint threat behavior and sham attacks.

Fig. 77. Defensive formation of a herd of musk-oxen against wolves. The adults form a ring with the horns pointing outward, the youngsters remain inside (after McBride 1971).

8.4.1.2. Search for Food

In contrast, living in groups can also be of advantage in obtaining food, e.g., for the hunter, and this again in two ways:

a. Combined Hunting. Many animals hunt in groups and enjoy a greater success. Lions in prides are, on the average, twice as successful as individuals. In the African wild dog, each pack member chases a gazelle closest to it. If the distance between them does not decrease, it observes the chase of a pack member and follows the more successful hunter. Fish-eating pelicans, mergansers, and anhingas circle their prey while swimming on the surface or drive the fish into shallow bays. In a similar manner, groups of killer whales and many other marine predators can encircle sea lions and walruses.

b. Exchange of Information. Another advantage, also of advantage for nonpredatory animals and for those species in which individuals forage alone, is the exchange of information. There are indications that the nocturnal roosting places of many bird species not only afford increased protection but are actual centers of information. Those animals that on the previous day have exploited a plentiful food source fly off in that direction the next morning directly and without hesitation—"pulling along" the undecided individuals. In primate groups, the exchange of information also plays a great role, especially when water and food sources are scarce, as in times of drought where much individual experience is required to find them.

8.4.1.3. Reproduction

Social living can also have biological advantages for reproduction. These are again in the area of predator avoidance. Thus, colony-nesting birds are able to drive off a predator more successfully than a singly breeding pair. This is possible even if each pair in the colony defends more or less only its own immediate nest site. Additional advantages are the attraction of females by communally courting males (see Section 8.2.2.1), the mutual stimulation and synchronization of pairs at the beginning of the breeding season (see Section 8.2.2.4), or the joint care of young. It is known in mice that the young grow faster when several females care for their young together. In a large number of animals, especially in colonial insects and in some mammals and birds, larvae or young are tended not only by their own mother or parents but by other members of the colony, group, or family (see also Section 8.4.3.1). Although the exact role of these "helpers" has not been determined in all cases, they contribute no doubt to an increased reproductive success. In more closely studied species, the number of successfully raised young was greater in families with helpers than in those where only the parents cared for the young.

8.4.1.4. Structures

Group-living animals can build structures that would be beyond the capability of individuals. This includes the structures of colonial insects, the

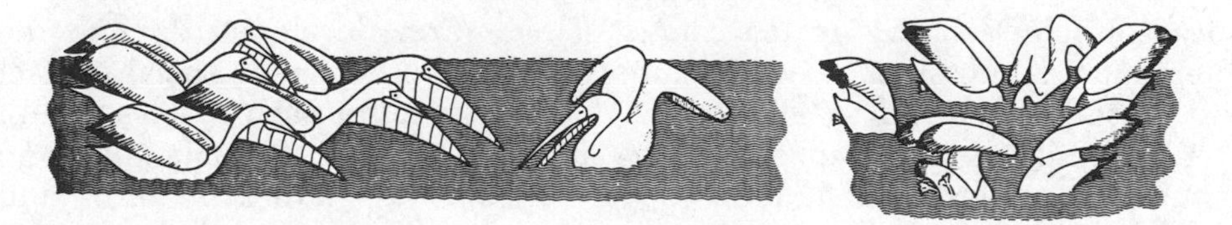

Fig. 78. Eastern white pelicans frequently hunt in groups and circle their prey (after Din and Eltringham 1974).

dams of beavers, and the giant communal nests of the South American green parrots and the African social weavers. The beaver dams, which affect the surrounding habitat, make possible the retention of a constant water level. As a result, the entrance to their dens is always protected under water. In the chambers of the social weavers, in which the birds also sleep at night, the temperatures are up to 23° C higher than outside. Without this the birds could not survive the cold nights during the winter.

8.4.1.5. Division of Labor

Within a social group, individual activities can be distributed in various ways among its members. This relieves individual animals of the obligation to perform all the behavior patterns that are necessary for their survival. This division of labor is especially developed in colonial insects. Here, the activity of an individual depends on its body structure, which can vary greatly depending on the "caste," sex, or age of the individual. In the latter instance, which is found in the honeybee, the same animal performs different roles during the course of its life. In the vertebrates, division of labor, e.g., leadership or guarding, is mainly found in group-living mammals. In some vertebrates that care for their young, there is a division of labor between the sexes, e.g., between guarding and feeding of young, or between collecting the food and distributing it.

This listing, which focuses only on some especially important functional systems, shows the many advantages that social living can bring for the individual group member. A group can always accomplish more than the same number of individual animals. Hence, many animals are obligatorily social, and they have either little or no chance of survival alone, as in the colonial insects. Frequently, several of the above-mentioned functions of a group are carried out simultaneously, and it is difficult to decide under which selection pressure the evolution of groups originally began, and which biological advantages were added secondarily.

8.4.2. Social Structure

The advantages of social living are offset by the disadvantages of intraspecific competition, especially for food, which can jeopardize the food base of a large number of individuals found close together in a small area. The need for camouflage is also in conflict with the aggregation of individuals in one place. Hence, the social structure of a particular species is always a compromise solution that during the course of evolution developed through the interaction of pressures between tendencies toward increased social contact and those that worked against it. It is characteristic for each species within certain limits, and it can show age- and sex-specific, annual, and even geographic differences. In most recent times, field studies on birds and mammals, employing increasingly sophisticated methods of investigation, have been able to show various and specific social structures adapted to ecological conditions. The results of these investigations permit the cautious generalization that species that live in the jungle, and whose food supply is spatially and annually distributed evenly, are found more frequently in pairs or small groups. On the other hand, species living in groups with many individuals live in more open and drier areas in which the food is available in large amounts at certain times or places but is scarce at other times or locations. However, there are many exceptions to this rule, and the social order of a species can only be understood after a detailed analysis of the circumstances in which it lives.

8.4.3. Altruism

8.4.3.1. Occurrence

As a rule, the social structure of animals is merely a "living together." A true living "for one another" is rare. In most species, members of a group do the same things they would be doing if they lived alone, and the biological advantages of group living are merely "side effects" that resulted when several species members did the same thing at the same time. A fishing pelican, a courting flamingo, or a bird flying from the roosting place to a rich food source behaves no differently in the group from the way it would if it were alone or in a pair, but the beneficial effect of its behavior for itself and/or for its conspecifics is greater in the group.

True ALTRUISM, i.e., a behavior in which an animal helps another to the detriment of its own well-being, as a rule is found only in the functional system of care of young. Between adult animals, i.e., between members of social groups, altruistic behavior is much rarer. However, the number of relevant observations that demonstrate this has greatly increased in recent times. This occurrence is found especially in three, although not always distinct, functional systems: mutual warning and defense, helping in caring for young, and exchange of food.

In common defense, altruistic elements are always then present when—as in baboons—individuals position themselves between the predator and other group members. In many social insects, defense of the colony is primarily or exclusively the task of specific individuals that are even morphologically distinct from other "castes" and that are frequently called guards or soldiers. Altruistic elements can also occur in behavior toward injured and sick individuals, although clear examples are few. Dolphins lift injured conspecifics to the water surface so that they can breathe (see Figure 79), and sperm whales and walruses circle injured conspecifics. Elephants try to get them to stand up, and dwarf mongooses keep sick group members warm and feed them and even let them eat first. Many similar observations need to be verified.

In many mammals and birds, group or family members help the mother or parents in caring for their young. In the European moorhen and various tropical and subtropical bird species of various taxonomic groups, the young of the previous year or of an earlier brood serve as "helpers." In some primate

Fig. 79. A group of dolphins lifts a wounded conspecific to the water surface (after Pilleri and Knuckey 1969).

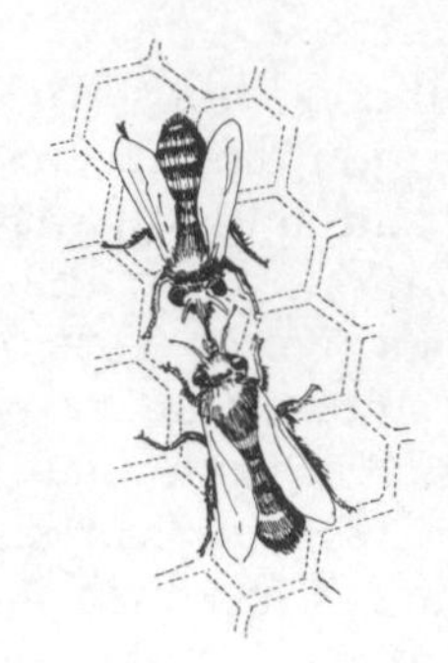

Fig. 80. Mouth-to-mouth feeding in honeybees (after Haas 1965 from Wickler 1967).

species, childless group members may temporarily take over young from their mothers, and in colonial insects, finally, the care of the young has completely shifted from the mother or parents to the older sisters.

Mutual food exchange between adults must also be considered altruistic behavior. It has been experimentally demonstrated in the honeybee. By feeding them radioactive substances, it was shown that there is a food distribution according to need within the hive. Those who have food give to those who demand it (see Figure 80). By these food transfers, the animals who cannot collect food, or not enough food, because they are engaged in other activities as part of the overall division of labor, are also fed. A similar system of food distribution is known in the African wild dog. In several higher primate species, especially in gibbons and chimpanzees, food exchange has been observed.

8.4.3.2. Biological Significance

The understanding of altruistic behavior in animals and its interpretation in the framework of a comparative phylogenetic view within ethology have until recently presented difficulties. This is understandable, since the origin of such behavior patterns could apparently not be brought into agreement with the idea of natural selection (see Chapter 10), which should favor only selfish behavior. New results and considerations have now closed this gap in our understanding.

This new understanding exists at two levels: (1) It has been found that various "altruistic" behaviors help not only the recipient but also the performing individual. (2) Population considerations suggest that pure altruistic behavior can be favored by selection under certain conditions of relatedness between the individuals involved.

a. Altruism and Selfishness. A number of frequently mentioned altruistic behavior patterns have upon closer examination turned out to be less than altruistic. The three following examples illustrate this. For one of the simplest forms of altruism, mutual warning involving warning calls and other signals, one can assume almost with certainty that fellow group members are informed about some danger. However, it is just as possible to think that the warner obtains an advantage, because as a result of mutual warning the success of a predator would be reduced, and it may leave the area.

In a similar way, social grooming or preening (see Section 8.2.5) at first benefits the passive partner by, for example, freeing him from parasites. On the other hand, an advantage for the active partner cannot be excluded: Most species that practice social grooming are contact animals (see Section 8.1.6). They sleep together, frequently in common structures or sleeping nests. Because of the always ever-present possibility of acquiring parasites, the freedom from them in the partner becomes as important as one's own.

Finally, there may be advantages for the "helpers" in caring for young. They can become experienced in caring for young before they have their own. Remaining in the familiar territory of their parents most likely has positive effects, and other advantages can be readily contemplated.

b. Altruism and Relatedness. In spite of the fact that the proportion of altruistic elements of the total behavior is not known in many instances, there can be no doubt that such behavior exists in animals. As with all phenomena among living things, there must be some selection pressure that influenced the development of the behavior patterns in question. Thus, we may ask whether or not altruistic behavior can have a biological function and what it might be. This question can be answered positively and with precision to the degree that the advantage of altruistic behavior is greater the more closely the recipient of the aid, i.e., the passive partner, is related to the active partner.

This leads to the following considerations: Natural selection affects phenotypes whose development is influenced by the genotype and the environment. The "success" of the underlying genotype is greater the more frequently the genetic factors of which it is composed are present in the next and following generations. This share of a population's gene reservoir is dependent upon the number of surviving and successfully reproducing descendants, and from the relatedness of members within groups where altruistic behavior occurs. The descendants of the first daughter generation have one-half, the grandchildren one-quarter, and the great-grandchildren one-eighth of the genetic factors of the original animal. In the same way, the genome of siblings is related by one-half (it is one in identical twins), by one-quarter in half siblings, by one-eighth in the case of cousins. From these genetic relationships one can obtain a measure of the "usefulness" of altruistic behavior. Even in the case of a complete sacrifice of an individual, the relative proportion of the animal's own genetic contribution would be the same in the next or the second generation if the consequence of the sacrifice were the survival of two direct descendants, four grandchildren, or eight great-grandchildren, which in turn would leave their own descendants at the appropriate reproductive rate.

Because of this dependence on the degree of relatedness, it is understandable that altruistic behavior occurs especially with respect to care of young because here, per occurrence of each behavior pattern, the survival of young is increased (by feeding them, giving warning, defending them, etc.). Hence, the rate of successful passing on and maximizing the potential of one's genes is greatest. However, even among highly developed social groups and societies, individual members, as a result of the origin of social groups from families (see Section 8.4.4.2), are so closely related that altruistic behavior could confer a biological benefit along the lines stated above. This is especially true for the societies of ants, bees, and wasps, in which the female workers are no longer able to pass on their own genes. Since the entire society consists of sisters, they can contribute to the survival of their genes by defending the nest and caring for the larvae of the sexually active animals and those produced by the queen(s). In line with these considerations, we are not surprised that altruistic behavior is especially developed in closed social groups (see Section 8.4.4.2). This behavior is in contrast to behavior leading to the actual killing—in extreme cases—of conspecifics at the level of intergroup interactions (see Section 8.1.8). Hence, altruism can be coupled with group selfishness.

These considerations, based on the degree of relatedness between individuals, should not obscure the fact that the advantages of altruistic behavior—as shown especially in studies of various bird species—can also be recognized in groups that consist of individuals that are not closely related. Hence, relatedness is not a precondition but merely an especially good opportunity for positive effects.

Overall altruistic behavior, quite in contrast to its assessment in human society, occupies no special position in the realm of social behavior, but evolved in the usual way under the pressure of biological necessities. The extent and form of altruism of a particular species are the result of a compromise that results from the often opposing demands of the individual's own survival and the passing on of its own genes to succeeding generations.

8.4.4. Types of Groups

The formation of groups among animals can be completely determined by the environment or it can, in addition, depend on social attraction, i.e., on a mutual attraction that brings conspecifics together and keeps them so. In the first instance, we speak of aggregations; in the second, of societies.

8.4.4.1. Aggregations

An AGGREGATION, sometimes called a COLLECTION, results when many individuals seek out the same locality, e.g., a water hole, a good hiding place, a suitable sleeping or wintering place. Here the animals do not seek each other out, but each comes on its own to the specific place. In a social sense their meeting is accidental. Such "pseudosocieties" are known in all taxonomic groups.

8.4.4.2. Groups

True GROUPS, which cannot be explained by environmental contingencies, occur in many forms. W. Wickler established a classification system and distinguishes between open and closed groups. OPEN GROUPS are characterized by the fact that their members are "exchangeable": the behavior of the entire group is not markedly disturbed or changed by the disappearance of individuals or by the arrival of newcomers. In CLOSED GROUPS such an exchange is not possible. Members of such groups relate to conspecifics essentially in two ways, each being treated quite differently: members of their own group, with which an individual has many social relationships; and members of other strange groups, toward which they behave either neutrally or aggressively, and which in exceptional instances may even be killed (see Section 8.1.8).

This polarity in behavior toward group members and strange conspecifics is not always clear-cut, since some exchange of individuals can also occur in closed groups. The extent to which a group is closed can, therefore, vary among individual species or even among various populations.

Both types of groups can be still further subdivided according to specific characteristics:

A. OPEN GROUPS
   1. *Open, anonymous groups.* These include swarms of insects, fish, and birds, and the migrating herds of mammals. In contrast to aggregations, these groups are usually not tied to a particular place. However, a clear distinction from a mere congregation of animals is not always possible.
   2. *Open, but not anonymous, groups.* A certain "advanced development" in a social sense is shown by the nesting and breeding colonies of birds and some fishes. They remain together for longer periods of time, and individual recognition can develop, at least between neighbors. However, here, too, the behavior of the colony is not affected by the disappearance of some and the arrival of other members, which settle on the vacated areas or along the periphery.

B. CLOSED GROUPS. Closed social systems require a mechanism for individual recognition that makes possible the distinction between group members and strangers. These can be divided as follows:
   1. *Anonymous, closed groups.* In these groups, which are called KINSHIP GROUPS, or SOCIETIES in insects, group members recognize one another by supraindividual characteristics. The function of such common characteristics is usually based on common odors. They can originate in two ways: odors produced within the body, or externally acquired odors. In many rodents, in peccaries, and in marsupial gliders, group members produce a uniform odor by rubbing and touching one another and marking with urine or scent gland secretions. In the honeybee, where each hive has its own odor, the specificity is achieved by the addition of odors from flowers that are preferentially visited at a particular time.

The significance of a group odor for mutual recognition can be readily demonstrated in an experiment: If one rubs a stranger with the odor of the group, he is accepted. If one covers a group member with a strange odor, he is treated like a stranger.

The advantage of such a communal odor for closed groups lies in the possible number of their members: while there are limits to individual recognition, group odors can be shared by any number of individuals. One honeybee hive can contain as many as 70,000 individuals.

2. *Individualized, closed groups.* In the most highly developed social organizations, known only in vertebrates, individual members know each other personally. It is only here that the disappearance of one individual can change the behavior of the group, and may even elicit searching behavior. Such individualized groups include troops of monkeys and apes and the packs of canid predators. In the African wild dog, which lives in packs made up of several males, females, and young, all adults participate in parental care. They "steal" each other's young to care for them, and the young are raised even when the natural mother has died. When the pack goes hunting, some adults remain as guards with the young. The hunters tear off pieces of meat, eat them, and bring them back to the young and regurgitate the food for them and the guards. Each hungry adult can beg from all other pack members. By this distribution mechanism, the fewer successful hunters can feed the remaining pack.

It must be pointed out that individual recognition of conspecifics, which is a prerequisite for the existence of individualized groups, is not limited to species with this social organization. This ability is known to exist in a large number of vertebrate species, especially birds and mammals, but also in some invertebrates, such as in some isopod and shrimp species. Individual recognition can include the partner, the parents, and the young, as well as territory neighbors, and pairs of nearby nests in the breeding colony. These may be recognized by sight, sound, or voice.

8.4.5. Origin of Groups

There are in principle two possibilities for the origin of groups: animals can come together or they can remain together. In the first instance, conspecifics, which have not lived in groups previously, seek out others at certain times for reproduction or migrations. In the second instance, conspecifics remain permanently together from birth on. The distribution of these two possibilities within the various kinds of groups referred to earlier is rather even. Open groups originate when animals come together for a special purpose, and they have evolved phylogenetically from aggregations. Closed groups, on the other hand, derived from the family and are based on the continued presence of adult young with their parents. This is true for the comparatively small kinship groups of mammals and birds to the same degree as for the colonies made up of thousands of insects. For these reasons we find highly developed social systems primarily in animal groups with extensive care of young. Among insects we find it only among ants, bees, and wasps and termites, two groups in which care of young is found widely distributed even in species that do not form colonies, such as in their ancestral species, e.g., cockroaches (see Section 8.3.2).

The reasons for the origin of closed groups from families are obviously related to two general characteristics of behavior involved in caring for young: there are always altruistic elements (see Section 8.4.3.1) and, as a rule, the behavior is directed from one to several conspecifics. Care of young hence contains excellent prerequisites for the development of bonding mechanisms (see Section 8.2.2.2).

Normally, a family is a social structure only for a limited time. Its biological function lies in raising young. After these are self-sufficient, the young leave their parents (see Section 8.1.9.6). In species that form closed groups, the immediate dissolution of the family does not take place. The time in which these longer-lasting families remain together can vary greatly. In some instances the separation from the parents occurs at the onset of sexual maturity of the young; in others the bonds may be maintained beyond that time, or may even last for life. This is true for insect societies as well as for many rodents, predators, hoofed animals, and primates.

8.4.6. Mechanisms of Group Formation

A functional system, which during pair formation and courtship (see Section 8.2.2.2) brings special problems to group-living animals and requires compromises, is intraspecific aggression. For reasons given earlier (see Section 8.1.2), aggression also must be maintained in social species, but it is directed only against members of strange groups. Within the group, aggression can affect the group's cohesion in a negative way. Under the influence of either positive or negative selection on intraspecific aggression, most social species evolved mechanisms that reduce aggression within the group and thus insure closeness between its members. These mechanisms include the establishment of a rank order and the development of behavior patterns with an appeasing function.

8.4.6.1. Rank Order

Closed groups are usually hierarchically organized. The structure of the hierarchy can differ from one species to the next. Sometimes the rank order is LINEAR, from the highest-ranking alpha animal, which is dominant over all others, to the lowest-ranking omega animal, which submits to or avoids all other group members. There are several variations from this linear rank order. Frequently there are TRIADIC relationships, in which beta is higher than gamma and gamma is above delta, but delta is higher than beta.

Rank orders can involve both sexes (e.g., in chickens), or there may be separate rank orders within each sex. The latter is true in males and females where the respective rank orders are independent, e.g., in rabbits, macaques, and baboons, or where the rank of the female is equal to that of the male with whom she forms a bond, e.g., in jackdaws. Such an "independent" rank is also known in young animals, whose rank in some species, e.g., in macaques, depends on that of the mother.

Rank orders can remain stable over long periods of time. As a rule, they show a certain dynamics where, at least among members close in rank, shifts may occur in which maturing animals gradually rise in rank and older ones drop. Some rank orders show periodic changes, as in baboon females, depending on the degree of estrus. Where the rank depends on that of the male, the female's rank can change instantly when a pair has been formed.

High-ranking animals have certain privileges such as eating or drinking first, the use of specific sleeping and resting places, and with respect to mating behavior. They may also have certain duties such as leadership tasks and guarding and defending the group. Frequently, high-ranking alpha animals interfere in the fights of lower-ranking group members, a behavior that is sometimes called PEACEMAKING. Finally, in some species high-ranking animals generally show a higher activity level than lower-ranking ones.

The phenomenon of social rank is, however, not exclusively limited to animals that live in groups. Species that live in pairs, especially when the bond is permanent, also show rank order differences. Usually the male is dominant. However, in some species there may be a temporary reversal of rank that protects the female, who is mainly responsible for the care of the young, against the aggression of a stronger male. Such a dominance change is known in especially aggressive species, such as the Indian shama thrush, but it is also found in many finches of Europe, e.g., bullfinch, chaffinch, and yellowhammer.

Rank orders are known in fishes, reptiles, birds, and mammals. In captivity there develops in many species a social hierarchy among those species that in the wild live a solitary life. Variable privileges, e.g., access to food—a carcass in the case of scavengers—or to water, can also be found between groups of social animals as well as between species. H. Hediger has given the name BIOLOGICAL RANK ORDER to the last example.

A rank order contributes to the stability of social relationships within the group. Conflict over social rank is usually restricted to the times when a rank order is established or when changes occur, such as when young animals mature. At all other times the various rights are generally accepted without conflict, or they are limited to fights between territorial neighbors. In this respect there is a certain similarity to the function of territorial behavior (see Section 8.1.5).

The existence of rank order was first discovered during observations on the domestic chicken, where rank order conflicts are carried out by pecking with the beak. For this reason the term PECK ORDER was originally used in ethology, but the term has increasingly fallen into disuse.

8.4.6.2. Behavior in the Service of Maintaining Groups

A second possibility for decreasing aggression within a group is the activation of behavioral tendencies that largely suppress attack and flight, and hence make possible a mutual approach and living together by group members. Behavior patterns that fulfill this function are called appeasement behavior or, since they occur especially when group members meet after some absence, greeting ceremonies (see Section 8.4.1).

Two primary "sources" exist for the phylogenetic origin of such bonding mechanisms. They already possess in their original biological function characteristics that inhibit aggression and facilitate bonds: parent–young relationships and sexual behavior. In fact, most behavior patterns that aid in the bonding of group members seem to be derived more or less clearly from these functional systems.

a. Newly Motivated Parental Care Behavior. Depending on the origin of closed groups from families, parental care behavior in the service of bonding within the group is widely distributed among species. In insect societies, adult animals feed each other mouth to mouth in the same manner in which they and some of their solitary ancestors fed the larvae with food. In vertebrates, too, behavior patterns of feeding, which already play a role in mating and pair maintenance behavior (see Section 8.2.5), are also in the service of group cohesion. The food-sharing of adult African wild dogs has already been mentioned. In Australian wood swallows, adult group members that are not paired also feed one another. Even in species where food is not actually passed, the form of the movement patterns indicates that the behavior was derived from the feeding patterns directed toward the young. Hence, group members often greet each other with movements that are identical with the begging movements of young animals, or patterns that are derived from this behavior. Examples are begging movements and calls in birds, pushing the nose at the corners of the mouth, and licking of lips in mammals.

Other components of parental care from which partial behavior patterns were taken over into the behavior inventory of the group include social grooming or preening and nest building. Hence, the mutual scratching, nibbling, licking, and searching through the fur—which in many species play an important role in pair formation (see Section 8.2.5)—occur also between unpaired group members. In some bird species, nest-building behavior patterns are used as greeting behavior in the context of social contact.

b. Newly Motivated Sexual Behavior. There is an additional bonding mechanism beyond the parental care behavior that in many species is the basis for the origin of bonding behavior within the group, and which evolved during phylogenesis. This mechanism, also found in solitary species, and hence in the ancestors of group-living species, exists between members of a pair. As a rule, this bond is maintained by behavior patterns that are part of mating behavior (see Section 8.2.5). Some of these behavior patterns, too, can serve to maintain social contact within a group. In some cichlid species a behavior that inhibits aggression within a group is identical with male courtship behavior. In the Australian masked grass finch, which belongs to the family of waxbills, group members greet each other with vibrating up-and-down movements of the tail, which corresponds to the female's invitation to mount. In many primates the invitation to mount, called PRESENTING, is used as a social appeasement gesture, which is used by lower-ranking group members who approach higher-ranking ones. In both examples, the behavior patterns in question, which in females are still a part of their sexual behavior, occur in both sexes in a social context.

8.4.7. Motivational Aspects

The biological advantages of groups, which were discussed in the beginning of this chapter, give us information about the selection pressures under which social organizations may have evolved during the course of phylogeny. The description of some of the most widespread social conventions shows us how social groups can exist in the present and how they are maintained.[13] What remains to be discussed are the motivations that underlie the formation of groups.

In principle, there are two possibilities: Group cohesion can be maintained by its own specific motivation—by its own "drive," as it were—or by the "detour" over specific bonding mechanisms that have been derived from other

[13] We find here the same two factors that are responsible, on the one hand, for the phylogenetic development of an adaptation and for those that regulate them in the present, as was outlined for annual rhythms (see Section 8.2.2.4). With respect to group formation, the ultimate causes include all those factors that contribute to an increased expectation of life in a group, such as increased protection against enemies and better opportunities to find food. The proximate causes consist of the sum total of bonding mechanisms that make possible living in groups today.

Fig. 81. Tail quivering, which in all waxbill species is a female invitation to mate, has evolved in the masked grass finch additionally into a social greeting between group members (after Immelmann 1962 from Wickler 1969).

functional systems, e.g., the sexual one. Concretely, the question arises whether, for example, the female invitation to mount, which was discussed above in the Australian masked grass finch and in primates and which no longer has a sexual meaning, is still sexually motivated or whether it is independent of it. As in the case of pair bonding (see Section 8.2.6), this question is hard to answer. Many observations indicate that with respect to group cohesion, there may be a specific motivation for it.

In the social groups of the Australian wood swallows there occur spontaneous assemblies several times a day both during and outside the breeding season. Some individuals land on a branch, the fork of a tree, or at a tree trunk, and all other birds in the vicinity stop all activity they were engaged in to join the other birds. This leads to strings or bunches of wood swallows even in the middle of the day under the hot noon sun. These aggregations are indistinguishable from the regular communal roosting behavior of these birds. The birds may separate again after 1 to 2 minutes and disperse in all directions, but they may remain together for as long as 15 to 20 minutes. Social behavior patterns such as begging, sitting close together with body contact, and preening can be observed with increased frequency. Environmental stimuli that could cause these aggregations have not been identified. They occur at all times of the day and in any kind of weather. They can only be interpreted as spontaneous discharges of a drive to seek close contact with conspecifics.

Similar observations have been described for other bird species, insects, fishes, and mammals. They indicate that the grouping of members may be based on a specific autochthonous social appetence that can be distinguished from other motivations. The individual behavior patterns that indicate this social motivation are part of this social drive in the same way as are behavior patterns in the sexual and care-of-young functional systems (see Section 2.5). In contrast to aggression (see Section 8.1.9.6), the biological function of a specific appetence to form groups, in view of its many advantages for the individual, can be readily understood. A conclusive proof for the presence of such a motivation has not been offered by anyone to date.

The concepts that have been postulated for a possible social motivation in the literature differ greatly. This is probably due to a lingering uncertainty about its true nature. Some authors speak of a social drive or, in mammals, of a herd drive ("instinct"); others avoid the controversial terms *drive* or *instinct*, and use neutral concepts such as *social tendency* or *gregariousness*.

9

Behavior Genetics

In the area of behavior genetics, scientists from two different disciplines are engaged in research. Geneticists use simple and easily recognizable behavior patterns to find answers to genetic questions. They are especially concerned with the difficult question of identifying genes for particular behavior elements. They work almost exclusively with fruit flies, especially *Drosophila melanogaster*, because the knowledge of the genetic makeup of a species for this kind of analysis is at present sufficient only in this species. These studies involve the elimination of single genes and examination of the resulting behavior changes in comparison to the behavior of untreated species members. Ethologists, on the other hand, focus on the behavior of the animals and use the methods of classical genetics, especially crossbreeding, to find explanations for the inherited bases of behavior. This is true especially with respect to the controversy about the inherited versus the acquired components of behavior (see Section 6.1.1), which can really be decided only in an individual instance by these kinds of experiments. A true bridge between these two areas of research has, unfortunately, not yet been built.

9.1. Methods

The number of studies taking the ethological approach to behavior genetics is still comparatively small. They progressed barely beyond simple crossing experiments. The reasons for this are primarily methodological: In order to obtain valid results—especially with respect to polyhybrids—large numbers of individuals are needed. Hence, these investigations are limited from the outset to species that can be kept under laboratory conditions, or that can be easily raised in enclosures and aviaries, and that produce succeeding generations in a relatively short time. Furthermore, these species should possess easily recognizable and measurable differences in their behavior, and they must also be closely enough related so that they can be crossbred. The resulting hybrids should produce fertile young, if one does not want to be limited to the F_1 generation. It is not surprising that all these conditions are met only in rare instances.

Behavior genetic investigations can in principle lead to the following results: They can lead to the identification of the behavior of hybrids between two species, subspecies, genera, or—in domesticated animals—races and breeds. These behaviors can then be quantified and compared with the behavior of the respective parents. Hence, they can give us a first glimpse into the inheritance of the behavior, especially with respect to the question as to whether or not the behavior of the hybrids is intermediate blending between that of the parents, or whether the behavior of one parent or the other is dominant (DOMINANT–RECESSIVE). By additional crossing with species' hybrids, one can examine whether a behavior pattern is controlled by only a single set of alleles (MONOHYBRID INHERITANCE) or whether several sets of alleles are involved (DI-, TRI-, POLYHYBRID INHERITANCE). Due to the methodological considerations discussed above, the second step has been carried out only a few times. Important conclusions about the inherited basis of behavior finally can be made only by artificial selection (see Section 11.3), i.e., through the selection of individuals with specific characteristics, and their continued breeding (see Section 8.1.9.2), preferably by inbreeding. An "experiment" that has already been in progress for hundreds of years is the process of domestication, where not only physical but also behavioral characteristics have gradually been changed by artificial selection (compare with Chapter 11).

9.2. Behavior of Hybrids

The behavior of hybrids has been studied in a large number of species from various animal groups—especially ducks and geese, pigeons, parrots, finches, and several species of fish and insects. In principle, they have all yielded similar results, which were expected from the laws of inheritance. In behavior patterns that differ only quantitatively in the two parental species, the behavior of hybrids is usually intermediate between the respective parents. Thus, male hybrids from the cross of a ring-necked pheasant and a domestic chicken, an example of a cross between two genera, show a posture when crowing that is exactly intermediate between those of the parent species. Hybrids between waxbills that drink by immersing their entire beak into the water and "pumping" the water, and those that drink by dipping their beaks, show an intermediate depth in the degree to which they immerse their beaks in the water.

If, on the other hand, the two parent species differ qualitatively, i.e., if one possesses a behavior that is absent in the other, then it is either present or absent in the hybrids, depending on whether or not it is dominant. In complex behavior patterns, hybrids may show a mosaiclike mixture of behavior elements derived from each parental species. This is found in a cross between two species of cichlids of the genus *Tilapia*, in which *tholloni* is a substrate breeder while *nilocita* is a mouthbreeder (see Section 8.3.1), which picks up the eggs in its mouth after spawning. Hybrids show random vacillation of the behavior patterns of both parents: they glue the eggs to the substrate, fan them, pick up a few of the eggs while continuing to fan the remainder, then again spit out the eggs, and so forth. Even if the eggs are fertile, the chance of successful raising of young is questionable from the beginning. A similar loss in the balance of the behavior has recently also been described in naturally

Fig. 82. The crowing posture of the ring-necked pheasant (left), a domestic rooster (right), and a hybrid of both species (center) (after Stadie 1968).

produced hybrids in two species of baboons, which also differ greatly in their social organization. The hybrids were observed on the borderline of the parental species' respective ranges. Their social structure consisted of a unique mixture of characteristics of each parental species.

Very informative results were obtained in crosses of two species of African lovebirds of the genus *Agapornis*. Some species of this genus, e.g., *A. roseicollis*, have a peculiar method of transporting nesting material to the nest cavity; they place strips of leaves or paper, which they have cut with their beaks, between the rump feathers on their backs, where they adhere to small hooks on the underside of the feathers. Other species, such as *A. fisherei*, transport the nest material in the normal fashion in their beaks. Hybrids between these two species at first try to carry the nest material between their rump feathers, but in this they fail: When they place the strips between the feathers, they fail to release them from their beaks, thus pulling them out again. They also do not push the strips far enough up between the feathers, or they try to fasten the strips at the wrong part of the body, e.g., the breast. These hybrids then show at first neither a true intermediary behavior nor that of one or the other parent species, although later on they learn to carry nest material in their beak. Instead we see here a disintegration of behavior elements that belong in one sequence.

Fig. 83. Birds of *Agapornis roseicollis* transport nesting material by pushing it between the feathers of the back and the base of the tail (after Dilger 1962).

These hybridization experiments with *Agapornis* have brought out two additional results: (1) The hybrids cut more and longer strips of nesting material than either parent species, a process that is called heterosis, which also occurs with respect to morphological characteristics; and (2) hybrids are often larger than their parents. In addition, attempts to insert nest materials in other parts of the body are the normal way of carrying the material in other species of the genus, e.g., *A. cana, A. taranta,* and are considered the original primitive means of carrying nest material in this genus. We apparently see a "throwback" to a behavior that is more similar to the ancestral form than to the behavior of the two parental species. Such "throwbacks" are known generally in the realm of morphology.

9.3. Crossbreeding Experiments

All the experiments cited thus far went no further than the F_1 generation because the hybrids were not fertile. The few instances in which experiments were carried into the F_2 generation yielded further understanding about the inheritance of behavior patterns. If one crosses two subspecies of phoretic[1] threadworms, *Rhabditis inerims,* a nematode—of which one subspecies makes an undulating movement with the anterior part of its body, increasing its chances of contact with an insect on which it will be carried away—then only "undulating" hybrids will be obtained. When one produces an F_2 generation from these hybrids, one obtains undulating and nonundulating animals in a ratio of 3:1. This shows that the character "undulating" is dominant and is determined by a single gene.

A dihybrid inheritance characterizes a behavior pattern of parental care that distinguishes two inbred lines of the honeybee. The behavior is the removal of larvae that have died in their brood chambers. The so-called hygienic bees uncap these brood chambers and remove the dead larvae. The nonhygienic bees lack this behavior. When both strains of bees are crossed, the resulting hybrids do not remove the larvae; they are nonhygienic. When this F_1 generation is backcrossed to the hygienic strain, four different results are obtained, of which two do not normally occur: In addition to hygienic and nonhygienic bees, there are animals that uncap the brood chambers but do not remove the dead larvae, and others that do not uncap the chambers but do remove the larvae if one opens the chambers for them. These behavior patterns occur in roughly the same frequency for each behavior. This indicates that the two partial behaviors are controlled by two separate genes and that both genes

[1] PHORESIA means the carrying of one organism by another without parasitism. It is found frequently in species that inhabit small and widely separated substrates and whose own means of locomotion are not sufficient to cross these distances. Thus, many dung-dwelling mites hitch a ride with dung beetles and are transported to a new dung heap. These nematodes, which live in excrement and decaying matter, hide under the wing covers of beetles and reach a new substratum in this manner.

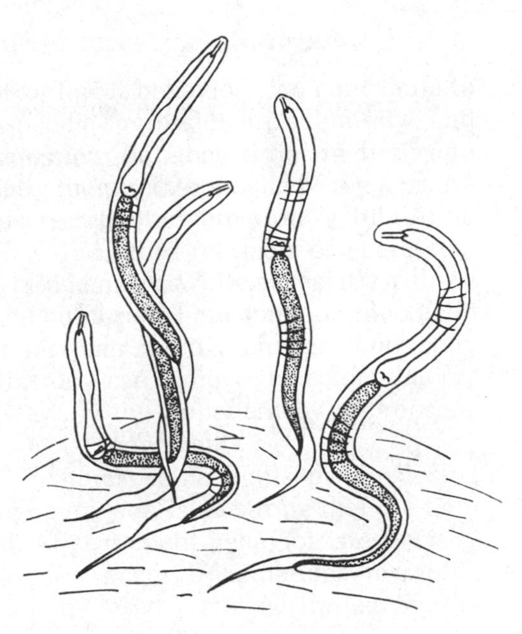

Fig. 84. Larvae of the threadworm *Rhabditis inermis* perform undulating movements with the anterior parts of their bodies, which are raised above the substrate (after Osche 1966).

are recessive. Complete hygienic behavior thus results only when both genes are present together.

All results known to date indicate that such mono- or dihybrid inheritance is the exception. Most behavior patterns seem to be polygenic. It is known from experiments involving the crossing of crickets that various components of their species-specific song patterns are independently inherited. The same is true for the wing vibrations of fruit flies with which the male stimulates the female during courtship. In the two species, the rhythm and extent of wing extension are different. Hybrids of two species of swordtail fish, *Xiphophorus helleri* and *montezumae*, show that even simple behavior differences in courtship behavior can be polygenic.The already mentioned mosaic mixing of behavior elements leads to the conclusion that they are controlled by separate genes.

Such polygenic control offers a great biological advantage over the all-or-none influence of a single gene in that natural selection can act upon gradations of a behavior, which allows for a gradual adaptation in small steps to changing environmental conditions (see Chapter 10).

It must be emphasized that we present here only examples for the "direct" effect of genes on behavior. (About the "detour" that must also be taken by direct genetic information, compare with Section 6.2.) Of course, genes can, as is true for the effects of hormones (see Section 5.2.3.2), influence behavior indirectly, e.g., by changes in the receptors, such as the sensitivity of certain sensory organs. The central nervous system, e.g., the capacity to retain memories, the production of hormones, the threshold of the hormonal response, or other morphological and physiological characteristics, all can be influenced by genes and thus affect behavior.

9.4. Conclusions

Currently available evidence indicates that:

- ☐ the genetic bases of most behavior patterns correspond to those of all other characteristics of an organism.
- ☐ most behavior patterns are controlled by many genes, i.e., they are polyhybrid.
- ☐ the behavior of an animal is a harmonic whole that can be substantially disrupted by the dropping-out of or changes in individual components, which have been observed in many hybrids.

These results have finally provided the evidence against all of those views and theories that deny a genetic basis of behavior, or that want to differentiate, in principle, the way behavior is determined from other characteristics of an organism.

10

Phylogenetic Development of Behavior

The variety of life forms, their function, and their way of life are the result of a continuously changing process that began with the origin of all life. It led to an increased unfolding of all living things, from simple precursors to the development of more highly organized and specialized forms. This process is called EVOLUTION. Its most important driving forces are mutation and selection. MUTATIONS are discrete "jumps"—changes in the genome. They are random events with respect to the direction of evolution and provide merely the raw material upon which natural selection can act.

SELECTION determines the survival and reproductive chances of an organism: individuals with advantageous characteristics that are best suited genetically to presently existing environmental conditions usually produce more viable and reproductively successful young than individuals with less advantageous adaptations. As a result, the incidence of certain advantageous inherited characteristics increases in a population, and that of less advantageous ones decreases. This results in a gradual change in the population in the direction of an optimal adaptation to the environment, i.e., it is "improved." In contrast to mutation, selection is always directed.

The behavior of a species is also subject to the laws of evolution. It developed from different precursors and can properly be understood only when its previous "history," its phylogenetic development, is known.

10.1. Methods

Methodologically such investigations are difficult, since one of the most important aids in the study of evolution—fossils—is, for all practical purposes, not available. While there are a few "behavioral fossils," e.g., petrified footprints, that enable us to conclude something about the method of locomotion of an extinct species, or petrified stomach contents and feeding tracks in sediment, which permit us to draw some conclusions about the manner of feeding, their extent and degree of certainty are so limited that it is generally impossible to draw final conclusions.

Statements about extinct species, which are so important for the reconstruction of the course of phylogenetic development, are not available in investigations on the phylogeny of behavior. References to the evolutionary history of behavior patterns are almost exclusively based on the behavior of recent or living species. There are two ways to study the evolution of behavior: the first is to look at the ontogeny of behavior, and the second is the comparison of closely related species, whose position in the zoological classification system and their probable "stage of development" within the particular group are already known from the data compiled in other biological disciplines.

10.2. Problems of Homology

In phylogenetic investigations of behavior, it is important to know whether the characteristics to be studied and compared are homologous or analogous. Here, ethology again encounters some difficulties not found in other areas, with respect both to methodology and to definitions.

HOMOLOGY is generally understood to be an agreement in structure and other physical characteristics, which depend on the inheritance from common ancestors. In contrast, ANALOGIES are similarities that evolved in various species independently of any phylogenetic relationships. Well-known examples for analogous structures are the wings of birds and insects and the lenses of eyes in vertebrates and cephalopods.

The main criterion for a homology is the existence of a common source of information—normally the genome—that directs the development of the particular characteristics. It is different with behavior, where we know of a second means of storage—memory. In behavior, this has consequences that can allow the same characteristic to occur in two species that are *not* closely related, as long as one species has acquired the behavior from the other. This means that carriers of homologous behaviors, in contrast to all others, need not necessarily have descended from a common ancestor. Hence, a homology in and of itself is not a criterion for phylogenetic relationship. For example, if one bird species learns the song or calls of another (see Section 7.4.5), then the vocalizations are homologous with respect to each other, even when both species belong to very different unrelated groups. Hence, we must clearly distinguish in ethology between a phylogeny of characteristics and a phylogeny of a group of animals. The former deals with the phyogenetic development and relatedness of characteristics; the latter deals with the phylogeny of the "cue-bearers," i.e., the animal species that have the particular character in question.

The way in which information is passed from one generation to the next also differs for both categories of homologous characters: INBORN characteristics are passed on by inheritance, and ACQUIRED ones by tradition (see Section 7.5). Of special interest for phylogeny are the so-called tradition homologies, where the learning of nonspecies vocalizations is an actual component of the learning program of a species. This is in instances where the brood-parasitic wydah birds learn the vocalizations of their host species (see Section 7.4.7.2).

Methodologically, the absence of fossils is a disadvantage for the study of homologies. Homologies must be laboriously identified by many characteristics and regularities. A number of "homology criteria" have been established that support the existence of a homology but cannot actually prove it. A homology can be assumed to exist when the occurrence of a behavior pattern or sequence agrees in as many individual components as possible in two different species. This is the criterion of SPECIFIC QUALITY. Additional evidence is provided when a behavior pattern occupies the same relative position in the functional organization of drives, e.g., in the same position in a behavior sequence: this is the criterion of POSITION. It is even possible that behavior patterns that are not too similar can be homologous in two species if there are graded transitions between them. Such intermediate forms can occur either in the ontogeny of the respective species or in closely related species. This is the criterion of LINKAGE BY INTERMEDIATE FORMS. The larger the number of these criteria, the greater the probability that a true homology exists.

10.3. Species Comparisons

If we compare the behavior inventories of related species, we can reconstruct the course of the evolution of individual behavior patterns fairly accurately. Sometimes one can arrange entire sequences of behavior elements of increasing similarity in a related group of species. In some instances, recognition of the origin of a behavior pattern in more highly developed species is possible only by comparing it with its expression in a more primitive species. Thus, many males of several waxbill species perform specific movements during courtship in which they hold a grass stem in their beak, which

Fig. 85. In many waxbills, as in the St. Helena waxbill (right), the male holds a feather in his beak during courtship. Other species, like the zebra finch (top), court without such a "nesting symbol" (see Section 8.2.2.2). Here the male merely hops toward the female with partially spread plumage and while singing. However, young males of this species occasionally show this behavior rudiment while actually holding a grass stem in their beaks (right-hand illustration after Steiner 1955).

clearly shows its origin in nest-building behavior. This use of the elements of nest building was "available," so to speak, to the waxbill group because in this group, in contrast to the majority of other songbirds, the male also participates in nest building. Some species that, based on many other characteristics, had been classified with the more primitive waxbills carry the same kinds of grass stems during courtship, and they are also used in the actual building of the nest. In some instances the grass stem may actually be incorporated into the nest at the end of the courtship. However, the Australian crimson finch uses a different material during courtship from that used when building the nest. Here, the grass stem is only a symbol. In several species the use of a nesting symbol has independently been lost. This development occurs in several intermediate steps: In some species a grass stem is used only during the initial phases of courtship but not during the actual courtship dance. In other species its use is optional, i.e., a male may court with or without a stem. In some species courtship dances with grass stems are only a rare exception. Even in species that no longer have this kind of courtship, the form of the courtship behavior clearly indicates its derivation from nest-building behavior.

10.4. Studies in the Ontogeny of Behavior

Findings of special importance for the phylogenetic development of behavior come from ontogenetic studies where, in the behavior of young animals, characteristics that appear to be similar or equal to the behavior of a more primitive species are found—as precursors, as it were, of the species.[1]

Thus, even in waxbill species that normally no longer court with a grass stem in their beak, as in the zebra finch, very young males occasionally still use a grass stem—a behavior that is again soon lost. The ground-nesting larks, which as adults run by alternating their legs, still hop with both legs together a few days after leaving their nest. This is the behavior found in closely related songbirds that nest in bushes. South African fur seals "stalk" in play, showing a behavior that, in their probably predatory seal ancestors, occurred most likely as prey-catching behavior in adults. The young of the robber crab, a tropical crab species that has taken to living on land temporarily, occupy empty snail shells, a behavior that is typical for the closely related hermit crabs in the adult stage. In these and similar instances it seems justified to draw some tentative conclusions about the behavior patterns of their ancestors.

[1] A similar recapitulation of phylogenetically older characteristics during ontogeny is known in morphology. The most frequently cited example is the appearance of gills, seen during embryonic development of mammals. This was first described by Ernst Haeckel in 1866 and has since been known as HAECKEL'S LAW, RECAPITULATION THEORY, or the BIOGENETIC LAW. The validity of a general statement on this phenomenon has been repeatedly disputed, and misunderstandings have also played a role. It is, of course, not true that ontogeny is an exact replication of adult characteristics of the respective ancestors. Instead, we are dealing with a greater similarity of embryonal and juvenile characteristics in the more primitive species to those in the more advanced species, because these characteristics are phylogenetically more conservative than those of fully developed adult organisms. This is also true for behavior.

10.5. Behavioral Rudiments

Behavior patterns that were carried along as "yesterday's adaptations" in their ancestors, although they meanwhile have lost their function—or at least their primary one—have been called behavioral rudiments. The examples thus far presented referred only to young animals in which such phylogenetic rudiments are most frequently seen. However, they also occurred in a few instances in adults, especially in cases where a movement pattern survived the body structure to which it "belonged" and with which it fulfilled its biological function. Many monkeys (macaque species) perform balancing movements with their tail, although they have only a short stump that is totally inadequate to fulfill this function. Many primitive deer, e.g., the muntjacs, bare their lips during threats and expose their elongated canine teeth (see Section 8.1.3). The same threat behavior is also shown in more highly evolved deer, e.g., the European red deer, whose canine teeth, which are no longer used in fighting, have become reduced as their antlers have evolved into larger structures. Hence, these animals threaten by showing "weapons" that they no longer have. Tree-nesting rails show typical egg-rolling behavior in experiments (see Section 3.9.2), although they are no longer biologically important because an egg, lost from the nest, cannot be retrieved by this behavior. In the courtship of some marine birds, such as guillemots and blue-footed boobies, behavior patterns derived from nest building occur, although the birds no longer build nests but incubate their eggs on bare rocks.

A behavior pattern that is often described as a typical rudiment is the head scratching in some birds when the bird lowers one wing, raises the leg on the same side, and brings its foot over the wing in order to scratch its head. When this "laborious" movement pattern is performed by young birds, they frequently lose their balance. It can only be understood in the context of the body proportions of a four-footed animal that can only reach its head with the hind foot by passing over the more anteriorly placed front legs. In spite of this, the behavior occurs in a surprisingly stereotyped manner, even though the animals would be physically capable of scratching themselves "directly" from below. Many parrot species simply raise one foot up front to bring food into their beak. However, when they want to scratch a spot at the base of their beak, they again scratch over their lowered wing, although their foot ends up at about the same place as when feeding. Hence, one can conclude that this behavior is a rudiment inherited from the reptiles, the direct ancestors of birds. However, there are some arguments against this interpretation: in some species, the young scratch themselves "directly" from up front, and the adults over the shoulder, which is contrary to the recapitulation theory, unless, of course, we are dealing with yet another special adaptation of the young. Furthermore, the distribution of both ways of scratching is apparently random among the more advanced as well as primitive bird groups. Hence, we cannot know which is phylogenetically older. A final explanation as to the source of this behavior is hence not possible at this time.

10.6. Ritualization

Many regularities in phylogenetic development can be shown to exist, in comparing the behavior of species as well as in ontogenetic studies, which facilitate mutual understanding between individuals and are called EXPRESSIVE BEHAVIOR (see Section 3.4). Since, in the process of communication, an exchange of information is needed that is as effective and unmistakeable as possible (see Section 3.4), many behavior patterns, especially those from the functional systems of aggression and sexual behavior, have changed during the course of evolution to insure improved transmission of signals. This is called the process of RITUALIZATION because of its similarity to the development of rituals in human behavior.

Ritualized behavior patterns are distinguished from nonritualized behavior in a number of characteristics that shown many parallels in the most diverse species, and whose development in the service of better communication is easily recognized. We are dealing primarily with simplifications, exaggerations, and formalizations of movement sequences, and the fact that many of these behavior patterns occur in usually rhythmically repeated sequences. The signal value of a movement can be further improved by the development of conspicuous physical characteristics, e.g., color patterns.

EXAGGERATIONS can be observed in many courtship and threat behavior patterns that are carried out with a much greater expenditure of energy than is used for the same behavior under normal circumstances. Examples are the strutting of a courting turkey, his wings dragging on the ground, and the slow, stiff-legged gait with which red deer walk toward or parallel to each other prior to their tournament fights (see Section 8.1.7). Many intention movements become signals only through exaggeration.

The second frequent change is the FORMALIZATION of behavior: the form of a ritualized movement is usually quite stereotyped, whereas its unritualized form is quite variable. The behavior patterns of locomotion, care of body, or food getting depend, in their intensity and completeness, on the actual requirements of the specific situation. But in their ritualized form as social signals, they always occur with the same speed and intensity, the "typical intensity." Examples are the many preening behaviors that occur during courtship, or the sounds produced by woodpeckers with their beaks, which vary greatly when the birds search for food but, during the so-called "drumming," show a unique, species-specific rhythm.

In other instances, the signal function of a behavior is enhanced by continuous, rhythmic REPETITION. This is true especially for auditory signals, e.g., the calls of many insects and amphibians, the begging calls of young birds, and the courtship songs and calls of many bird species. The continuous cooing of a male pigeon comes to mind. Examples of stereotyped repetitions of visual signals are the waving of pincers by fiddler crabs and the flashing light signals of lightning bugs.

Ritualization is not only a phylogenetic process: it can also occur during ontogenesis, as illustrated by the development of stereotyped behavior patterns in adults from the more variable behavior of young animals. A good example is the change from the variable songs of young birds to the more or less uniform song of adult males in many bird species. This is called ontogenetic ritualization.

A formalization similar to that of ritualization of behavior patterns can occur in unnatural situations in adult animals. Many zoo animals develop stereotyped movement patterns due to a lack of adequate opportunity to exercise. Behavior of a rather compulsive nature—e.g., head turning, rocking movements, body turning, walking or running in fixed paths—all illustrate this kind of behavior. The begging movements of some zoo animals also seem to have a ritualized character.

Practically all functional systems are candidates as sources for ritualized behavior patterns. Most frequently they originate in intention movements and displacement activities.

Intention movements are incomplete movements, e.g., the pecking at nesting material before the actual beginning of nest building, or the conspicuous bending of the legs in the joints just prior to flying off, seen in many birds. The behavior occurs when the stimuli for the behavior patterns in question have not yet reached the necessary threshold (see Section 2.2). In their ritualized form, these behaviors are important social signals. Thus, many threat behaviors, e.g., baring of teeth, are intention movements of fighting; the courtship with grass stems in waxbill finches (see Section 10.3) can be derived

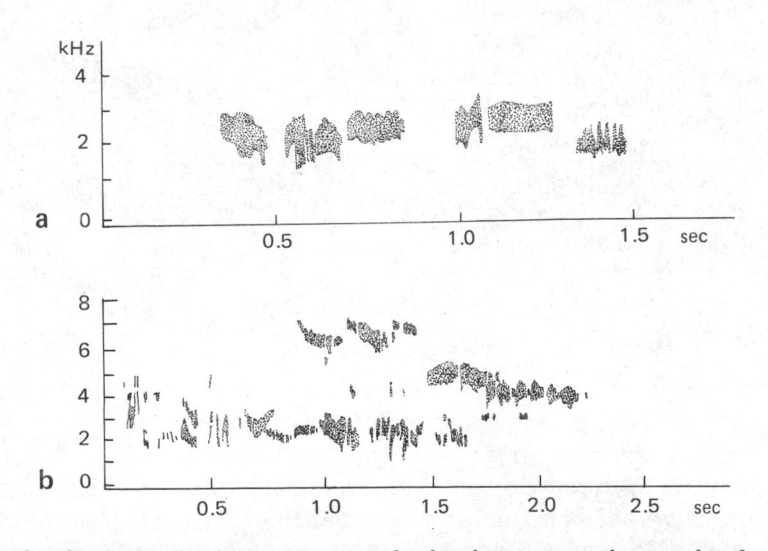

Fig. 86. The development of song in a mistle thrush is a typical example of ontogenetic ritualization: the stanza of the adult male consists of a few relatively uniform elements (a). The song of a young bird, the so-called subsong, is more variable, and various elements not only differ greatly but are also characterized by an overall broader frequency spectrum (b). In many bird species a subsong occurs also in adult males, especially at the beginning of a new reproductive period. Thus, in each year there is a new "ontogenetic ritualization" (after Thorpe and Pilcher 1958).

from intention movements of nest building; and the lip-smacking greeting of guenons is probably an invitation to social grooming. The bending of legs in flocking birds is especially exaggerated and seems to have a synchronizing effect on the takeoff of the flock (see above). It appears as though the birds mutually stimulate each other until they are finally motivated enough to take off.

Displacement behavior can also secondarily acquire a signal function in social communication. It has most frequently been incorporated into courtship and threat behavior. This is not surprising since in both functional systems there exist already conflicting motivations (see Sections 8.1.3 and 8.2.2.2), which result in many displacement behaviors. One of the first studied examples in this direction comes from preening behavior, the so-called sham preening in various species of ducks.

Besides a change in form and function, a change of motivation can occur in the process of ritualization: thus, the intention movement for nest building during courtship in waxbill finches is no longer associated with nest building but is now sexually motivated. The same is true of ritualized preening behaviors that have become an integral part of courtship.

10.7. Convergence of Behavior Patterns

The previous examples of phylogenetic development of behavior pertained to the problems of group phylogenesis and concentrated on those characteristics (see Section 10.2) that were most probably homologous in the sense of being phylogenetically related. Similar insights into the possibilities of evolution are available through the study of behavioral convergence.

First, we need to define the concept of *convergence* and differentiate it from homology. Homology refers to the existence of a common store of information, which insures the transmission of information from one generation to the next and thus acquires the adaptations before its members have encountered the particular environmental contingencies to which these characteristics are adapted (see Section 6.1.1). Hence, information about the environment is "indirectly acquired."

Fig. 87. Intention movement of takeoff in the wheatear (after Tinbergen 1952).

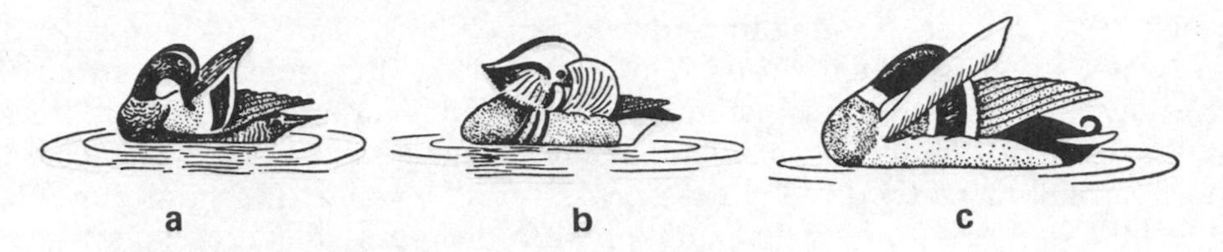

Fig. 88. Ritualized preening movements during courtship behavior in ducks, which originated in displacement preening—sham preening: (a) garganey drake, (b) mandarin drake, (c) mallard drake (after Lorenz 1941 from Tinbergen 1955).

With CONVERGENCE, the information is acquired *directly*, i.e., where certain characteristics have been acquired in individuals or species in similar or identical environmental conditions independently from each other. Since there is no common source of information, the adaptations become gradually more similar as they respond to environmental contingencies. The fins of land vertebrates that have secondarily returned to life in the water are morphological examples of convergence, e.g., in reptiles (ichthyosauruses), in birds (penguins), and among mammals (seals and whales). These structures are homologous in being limbs, but as "fins" they are analogous, and they became similar as a result of convergent evolution.

In behavior, convergences can evolve within as well as between two or more species (intraspecifically and interspecifically), and the characteristics may be innate as well as learned. When a young animal learns to avoid predators or what to eat by tradition (see Section 7.5), the behavior is homologous in all animals, and its development is independent of its own experience with the object. The animal "knows" from the reactions of its conspecifics whether a prey animal is edible or not without having tried eating it itself. If, on the other hand, each individual must acquire such experiences on its own, then the behavior is the result of convergence. The same is true between species: if two species possess the same behavior as a result of a close relationship, or because one has learned it from the other, it is homologous; if the behavior has been separately acquired, it is based on convergent development. An example of homology is the hovering flight of hummingbirds. However, the same hovering flight is convergent when one compares hummingbirds and the similar but less highly evolved hovering flight of the nectar birds and honey-eaters, because in these three groups the ability to hover evolved independently as an adaptation to feeding on flowers.

Convergent evolution always depends on the adaptations to similar environmental conditions. An example is that most ground-dwelling bird species walk, regardless of their systematic relationship to each other, while most tree-dwelling species hop. Egg rolling (see Section 3.9.2) evolved in ground-nesting birds in various groups, e.g., ostriches, ducks, chickens, wading birds, cranes, and others. It certainly seems to be an adaptation to the lack of true nests and hence the danger of eggs rolling out. Some Australian waxbill finches "pump" water in the manner of pigeons by immersing their beaks into the water, while others drink like chickens and dip their bills into the water and raise their heads. The comparison of species shows that the time-saving pumping has apparently evolved independently in several groups in species that live in arid habitats, whereas related species in biotopes with more rainfall drink by dipping their beaks into the water. The pumping mode of drinking seems to be an adaptation that decreases the time spent at the water holes in open country, where the birds are more exposed to predators. Another good example of convergent evolution of behavior is the already mentioned warning call (see Section 3.5) in small birds. These calls have become so similar across species that they can be considered an interspecific signal.

The investigation of convergences in behavior that correspond to a species' needs in a particular habitat is of general interest, since it permits general statements about the possibilities for adaptations and forms of expression in species or related species within a group, thus contributing to the study of evolution overall.

10.8. Conclusions

Current knowledge of the phylogeny of behavior permits us to draw three important general conclusions:

- ☐ The evolution of behavior is subject to the same laws as the phylogenesis of all other characteristics of an organism. Although this seems self-evident, it frequently has been doubted or denied.
- ☐ Based on this fact, homologous behavior characteristics, with the exception of homologies by tradition, have the same high taxonomic[2] value as morphological and physiological characteristics, i.e., they can be used in the classification of species, genera, and other systematic categories.

[2] TAXONOMY, OR SYSTEMATIC BIOLOGY, is a biological discipline that classifies organisms into a "natural" system that reflects the phylogenetic relationships of animals, making possible an overview of all the various and confusing forms of life.

Fig. 89. Zebra finch immersing its beak in the so-called "pumping" mode of drinking.

☐ The behavior can be of decisive importance in evolution in general, and it is considered to be in the forefront of evolutionary activity.

This last conclusion is without doubt the most important one. On the one hand, it refers to the fact that there is the possibility of passing on information by tradition from one generation to the next, which is not known anywhere else; but it also leads to the recognition that changes in behavior are often the first step in an evolutionary development that *later* can also bring about changes in other characteristics. Behavior, then, seems to be a kind of pacemaker in the process of evolution.

This applies especially to food preferences. A new preference or feeding strategy that is individually acquired, and that may be passed on by tradition, does not in and of itself constitute a new step in evolution, since it can disappear with the death of the individual or some other break in the tradition. However, it brings the animal that has the behavior into a new adaptive situation in which *other* selection pressures may exist, and in which other characteristics may be of advantage. This in turn may influence the success rate of the randomly occurring mutations, and hence the change in behavior may gradually have morphological consequences, e.g., changes in the structure of mouth organs in insects, the beaks of birds, or the stomach in mammals. These in turn may enable the animals to exploit a new ecological niche.[3]

The same applies to changes in the mode of locomotion: the "gliding jump" is probably the first step in the development of the ability to fly. It evolved independently in four classes of land vertebrates and reached various degrees of perfection. This change led to fundamental morphological and physiological changes of the entire organism.

Hence, it is possible that many, if not most, evolutionary changes were initiated by changes in behavior that thus became an important causal factor in evolution. As summarized by E. Mayr (1967), behavior can—like all other characteristics of an organism—evolve and, furthermore, it can also cause evolution.

[3] The term ECOLOGICAL NICHE refers to the role that a species occupies, based on its demands on the environment and its utilization of environmental conditions within the ecosystem. Hence, it refers to the relationships of organisms in a living community to each other and to their common environment. The term NICHE does not refer to a spatial entity but refers only to a reference system between various species and environmental contingencies.

11

Influence of Domestication on Behavior

11.1. Problems of Definition

A process that, like phylogenesis, leads to changes in species, and whose significance for a better understanding of the evolutionary process has been discussed in the literature, is the DOMESTICATION of animals. It is a "case study" of evolution as a whole. Animals and plants that have gone through this process are considered domesticated, and they differ from their wild ancestors in a number of structural, physiological, and ethological characteristics.

By definition, a characteristic is an example of domestication when the difference from the ancestral form is genetically determined and can thus be passed on to succeeding generations. However, it is frequently difficult to make a clear distinction: since the development of many structural and physiological characteristics also depends to a large degree on environmental factors, the artificial conditions in which the animals live can also lead to modifications that are superimposed on any possible genetic changes. They may have the appearance of such genetic changes when, in fact, they are merely adaptations to specific environmental conditions. We can see this in captured wild animals—especially when they are caught young. We mention here only the changes in the hormonal system, the decreased development of sense organs and parts of the central nervous system in environments impoverished of normally encountered stimuli.

In behavior, such environmentally induced modifications are even more difficult to discover and to distinguish from the genetically based results of the domestication process. (See Section 6.1.2 for a discussion of the problems in recognizing inherited behavior patterns.) Hence, the ethological analysis of domestication is still in its beginnings. Many generalizations made to date are most likely not generally valid. Nevertheless, this kind of research has engendered much interest in recent years, since the human species apparently has also undergone what is referred to as self-domestication. There is hope that results from investigations on the domestication of animals may shed some light on possible parallel phenomena in humans.

11.2. Characteristics of Domestication

The changes we see in domesticated animals in comparison to characteristics in the ancestral or wild form indicate two generally valid rules detected early in the study of domestication: On the one hand, domesticated animals show a greater intraspecific variability in structure, physiology, and ethological characteristics. Furthermore, there is a conspicuous "convergence" among many characteristics "typical" of domestication, and which can be seen even in species that are not closely related. In many domesticated species, we see the emergence of long wool or curls, piebald spotting, or shortened legs such as in dachshunds, swine, and sheep. The central nervous system, which is of special interest with respect to behavior, also shows parallel changes. There is a general reduction of brain weight and a decrease in differentiation in the phylogenetically most recently evolved parts of the brain. In some instances, this parallel in characteristics is so obvious that the conclusion is drawn that domestication-related changes may have their own laws.

The behavior of domesticated animals can also differ from that of their wild forms, and many parallels can be seen. Quite early Lorenz called attention to some consistent changes in behavior patterns that developed during the process of domestication. The most important of these are increases and decreases of the readiness to act (threshold levels), i.e., with respect to the action-specific energy (see Section 2.6) of a behavior pattern, as well as in the changes of innate releasing mechanisms.

11.2.1. Changes in the Readiness to Act

Some behavior patterns occur less or more frequently in domesticated animals than in the wild form. The first change, called HYPOTROPHY, applies to aggressive behavior; warning-, escape-, and defensive-behavior patterns; the behavior associated with care of young, e.g., "broodiness" in the domesticated chicken as compared with the jungle fowl; and even to the spontaneous locomotor activity level of an animal (see Section 5.2.3.4). In the extreme case, some behavior patterns can disappear altogether. Such ATROPHY was demonstrated in the song of very highly selected canaries, which lack several elements of the song of the wild serin, or where they occur only rarely.

In contrast, HYPERTROPHY is a change that occurs especially in sexual behavior, the so-called hypersexuality of domesticated animals. It is characterized by a general increase in sexual behavior, and a decreased dependence on particular seasons (see Section 8.2.2.4), up to complete independence of them. In sexual behavior there may be a relative increase of those behavior patterns that are associated with an especially high sexual motivation.

This last change may lead to shifts in the total sequence of courtship behavior in such a manner that the "initial" elements, which in wild animals constitute a substantial part of the entire sequence, may be temporally much reduced or be entirely eliminated. Hence, it consists only of copulation itself and a few of its immediate behavioral precursors (see Section 4.1). This differentiated frequency of partial elements of a sequence can lead to the performance of individual parts of a behavior sequence, e.g., certain prey-catching patterns in cats and dogs, without the appearance of normally preceding or succeeding elements.

11.2.2. Changes in Innate Releasing Mechanisms

Usually, wild animals react in their social behavior to very finely attuned stimulus configurations. In longer action sequences the occurrence of the next elements depends on the appropriate preceding releaser (see Section 4.1). This selectivity is usually greatly reduced in domesticated animals, especially for sexual behavior, but it also applies to care of young and other areas of social

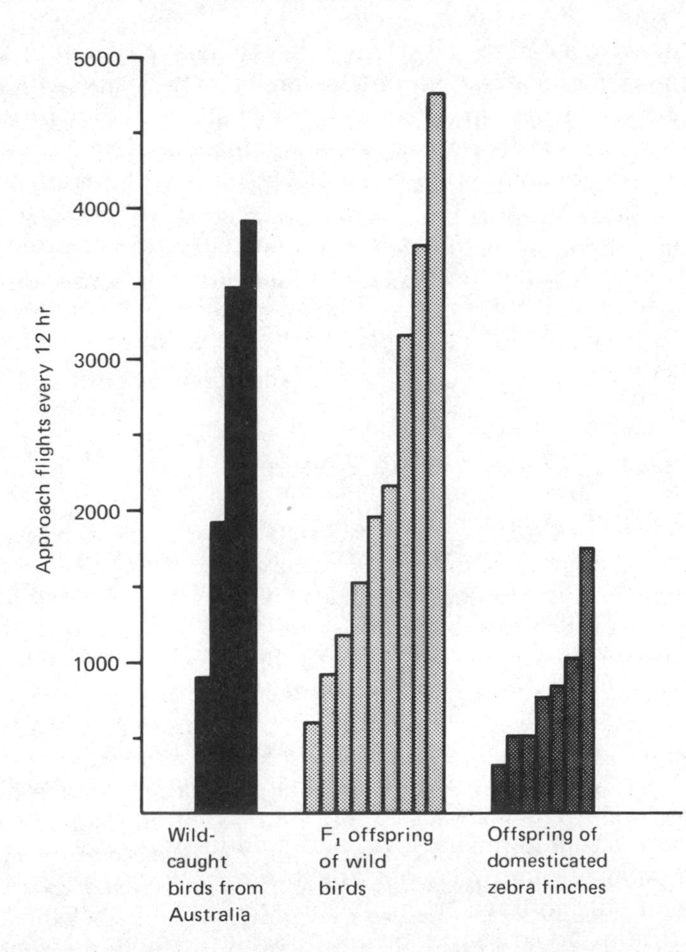

Fig. 90. The differences in activity in wild and domesticated zebra finches. The bars signify the number of automatically recorded flights from one perch to the other during a 12-hour light period. Each block represents one experimental animal. The animals in the second and third groups were raised under identical conditions (after Sossinka 1972).

behavior. For this reason, crosses and cross-fostering are usually easier in domesticated than in wild animals. Many jungle fowl, the wild form of the domestic chicken, lead only their own young, which they recognize by their distinctive head markings. Domestic hens, on the other hand, also brood the young of other species of fowl and even those of unrelated precocial birds, e.g., ducks. This varies among the different breeds of chickens. It must be pointed out that comparative, quantitative investigations on the extent of the effects of social experiences are not yet available. Only when they are will it be possible to determine whether in these cases innate releasing mechanisms are involved, or whether the decreased selectivity is the result of a decreased ability to learn.

There are also additional examples from almost all functional behavior systems. Learning ability is especially implicated, in that many domesticated animals show substantial decreases in this area.

In general, we can make the following statements about changes in behavior during domestication: No new behavior pattern has ever appeared, and in only a few cases has a behavior pattern existing in the wild form been lost. Even the sequence of innate behavior patterns, i.e., the form of the fixed action patterns (see Section 3.9.1), is usually preserved. In cases in which a change took place, it has always been a loss of differentiation, but never a "positive" new development. Hence, most of domestication-related behavior changes are quantitative and involve the frequency and selectivity changes already discussed. The process of domestication, then, is a good example for the extent of the overall effect, which can even bring about quantitative changes of such magnitude.

11.3. Causes of Changes during Domestication

The possible causes for the origin of characteristics of domestication are found first of all in an increase of genetic variability in comparison to the wild form, and a change in the direction and intensity of the selective forces that affect organisms. An increased mutation rate under artificial housing conditions was suspected, especially considering the speed with which they take place. The reasons for these hypotheses were already discussed earlier. Convincing evidence, however, has not yet been presented. On the other hand, the changes in selection pressure on domestic animals are substantial. Natural selection has been replaced by artificial selection, whose intensity is increased because of the relatively small number of individuals found in domestic populations. The direction of selection deviates greatly from natural selection and may actually run counter to it. Hence, in many, especially the large domestic species, the least aggressive individuals are selected for their greater ease in handling. The chance of survival in the wild for animals with these characteristics would probably have been less in the defense of territories and/or partner under natural conditions. The occurrence of reproductive behavior in wild animals outside of the species- and population-specific seasons would certainly be selected against because of the lower chances of survival. In domestic animals, there is selection for increased sexual behavior, because the breeder is interested in an increased rate of reproduction.

A number of examples show how strongly the changes in frequency and shifts in behavior patterns depend on the particular goal of selective breeding. Whereas strong aggressive tendencies are generally eliminated, they are selected for in animals that are used in fights, where their underlying motivation hypertrophies. This development in opposing directions can even occur in one species: examples are the various breeds of dogs, comparisons between East Asian fighting cocks and the cocks of Central European chicken breeds (see Section 8.1.9.2), and also the highly selected song canaries, which have lost those song elements of the wild form that are associated with the highest sexual motivation, in contrast to canaries that have undergone no particular selection for their song, and which sing those particular behavior elements more frequently. This agrees with the overall hypersexualization of domestic animals.

The parallel appearance of domestic characteristics is almost certainly, at least in part, the result of similar selection by the respective breeders of various animal species. The best example is the hypersexuality of domesticated animals (see Section 11.2), which results from selection for characteristics that increase the rate of production. This includes increasing independence from a limited reproductive season and an increase in the speed of development and earlier sexual maturity. This parallel development owes its dependence to artificial selection. This becomes obvious when one compares the classic domestic animals, which were derived from wild forms that have a regular and temporarily limited reproductive season, with a very "young" domesticated animal like the Australian zebra finch. These birds have evolved adaptations to the arid conditions in north and central Australia, which in various populations fit the irregular climate. They have a number of physiological and behavioral peculiarities that allow a rapid onset of breeding and a high reproductive rate as soon as rains create favorable environmental conditions. These adaptations include an early onset of sexual maturity in comparison to related comparable species. Hence, there is, under natural conditions, already a strong selection pressure for early sexual maturity, which insures that at any time a large percentage of the population is ready to reproduce. This selection pressure has not increased in captivity. On the contrary, it has been decreased compared to that in the wild, since most breeders want to avoid a weakening of their breeding stock and hold off breeding the birds until they are older than half a year. Consequently, the young of domesticated zebra finches show slower sexual development in comparison to the wild form, a development that runs counter to what is found in most other domesticated animals.

The reason for the rapidity of artificial selection, which has also given rise to a suspected increase in the mutation rate, is readily seen: artificial selection is distinguished from natural selection by its rigor and the rapid changes in directions of selection, which are not found in nature. Furthermore, this selection is usually applied toward only one characteristic, although, because of the pleiotropic effects of genes, many other characteristics are unwittingly affected. Because of the small size and frequently extreme isolation of many domestic animal populations, the probability is larger that recessive genes express themselves, and this can lead to a further increase in the rate of change.

11.4. Conclusions

The differences that exist in the behavior of domestic animals and their wild ancestors, and the probable regularities in the behavior changes that occur during domestication, should not obscure the fact that it is difficult—because of problems of definition and delineation—to discover the extent of domestication. The difficulty in recognizing its genetic basis has already been discussed. Thus, it is often hard to decide whether an observed change in the readiness to act is truly inborn or whether it depends on changed environmental conditions, e.g., being kept singly. Because of the long-term influences of early experience, especially with respect to motivation (see Section 7.4.7.2), great stability in behavior can be observed even in the absence of a genetic basis, and one might wrongly suspect innateness. Similarly, the decreased selectivity in the responsiveness to social stimuli can be the result of a lowering of threshold (see Section 2.2) and/or a modification of a readiness to behave a certain way as a result of early experience. Hence, all conclusions about the effects of domestication require great caution, given the present state of our knowledge. Additional insights can only come from studies in which domesticated animals and their wild forms are raised and maintained under identical environmental conditions.

We must be even more careful when discussing possible parallel phenomena in humans, since we need to consider the additional difficulties in distinguishing between man's self-domestication and the domestication of animals. In domestic animals, natural selection has been replaced by an artificial selection that can be as rigorous as, if not more so than, natural selection. In humans such a shift in selective pressures has not taken place; instead, it is characterized by an absence of, or at least a general decrease in, the effectiveness of numerous natural selection pressures.

12

Ethology and Psychology

12.1. Introduction

Since the beginnings of scientific ethology, the question was posed concerning to what extent the knowledge and insight gained from the study of animals could be applied to an understanding of human behavior and psychology. This discussion always involved speculations and was laden with misunderstandings. Opinions differ, not only outside ethology but also within the discipline. Response ranges from the incorporation of ethological results and theories into psychology to the total rejection of any attempts at such comparisons. The first position is based on the unquestionable fact of our human origins from nonhuman ancestors, and on the agreements and similarities of physical characteristics and functions that follow from this acceptance. It is hard to imagine why behavior in its totality should be an exception to this. The opposing opinion is rooted in the view that humans have diverged from all nonhuman organisms, especially with respect to behavior, so that comparisons are not valid—at least there are no theoretical or practical consequences. There may, of course, be philosophical or other reservations as well.

12.2. Level of Comparison

One of the reasons for the various attitudes toward ethology are the different levels on which discussions about the applicability of ethological knowledge take place. The following basic premise needs to be stated: It is true that no results obtained from one species can be applied to humans. By the same rule, it is not permissible to do this from one animal species to the next becase of the uniqueness of the behavior inventory of each species. However, ethology as a comparative science has a great methodological advantage over human psychology: it can investigate several species or larger taxonomic groups (see Section 1.2). From this comparison certain generalities in the development of some functional behavior systems can be discovered that are characteristic for particular groups of animals. Such a comparison, which begins at the level of individual, and closely related, species, can be extended to larger systematic groups, e.g., families, orders, and classes, provided enough information is available. For example, one can find common behavior characteristics among Old World primates, among primates in general, and, even more broadly, among all mammals.

At this level of abstraction and generality, of the groups of organisms that include humans in a biological sense, e.g., primates and mammals, such comparisons seem, in principle, possible and acceptable. However, even here no direct extrapolation can be made. Rather, the knowledge gained from the ethological study of animals should always be considered merely as working hypotheses that may stimulate human psychologists in the investigation of possible parallel phenomena in human behavior.

In addition to these limitations imposed by the level of comparison, there are differences with respect to the various areas or functional systems in human behavior and hence in the various subcategories of human psychology. Any possible agreement is more probable in those behavior characteristics that are closely connected to their physiological basis. Furthermore, we can assume that the chances of discovering similarities are greater in the early stages of behavior development than in the fully matured adult organism. For this reason, the search for possible comparisons is most promising in the areas of physiological psychology (sensory physiology, neurology), clinical psychology (rehabilitation, drug research), developmental psychology, and learning psychology. Other areas of human psychology deal with behavior that is "typically human," i.e., behavior that is farthest removed from its biological basis, or where such a basis is no longer recognizable; therefore, a comparison shows little promise. This applies especially to personality—and to organizational psychology, to various areas of psychoanalysis and testing psychology, and to most linguistics.

12.3. Application

Once it is recognized that animal–human comparisons are possible, keeping in mind all the limitations, then we can consider not only the theoretical advances and insights but also the practical consequences of this approach. These are in the quantitative and experimental investigations on *animals,* and in the application of ethological research methodology to the study of *humans.* In the first instance, by taking advantage of animals as research subjects, problems can be studied and questions posed, the answers to which can lead to a better understanding of human behavior—especially in situations where ethical considerations prohibit similar research on humans directly.

12.3.1. Animal Experiments as a Model

An impressive example of animal research in lieu of humans is the investigation of parent–young bonds in primates. This began with the work of H. F. Harlow and his co-workers on rhesus monkeys. The work has since been continued on various other species as well. Harlow separated young rhesus monkeys from their mothers at various ages and for varying periods of time and raised them on cloth-and-wire substitute mothers (see Figure 91). The results of these and subsequent experiments led to these general conclusions:

☐ The presence of a close mother–child bond is a prerequisite for later socializing processes.
☐ The consequences of a temporary separation from the mother vary in severity, depending on the age at which they occur.
☐ The establishment of a bond is due not so much to the "reward" of food (milk) as to the contact comfort provided by the mother's body.

Comparative studies have confirmed that these principles are widely applicable in primates, although the time of the sensitive age periods for these effects varies. They remind us of the sensitive phases in imprinting processes (see Section 7.4.7.1), and the extent of the separation trauma may differ from one species to the next.

These studies on nonhuman primates were not without influence on developmental, psychological studies of humans. Many parallels were found

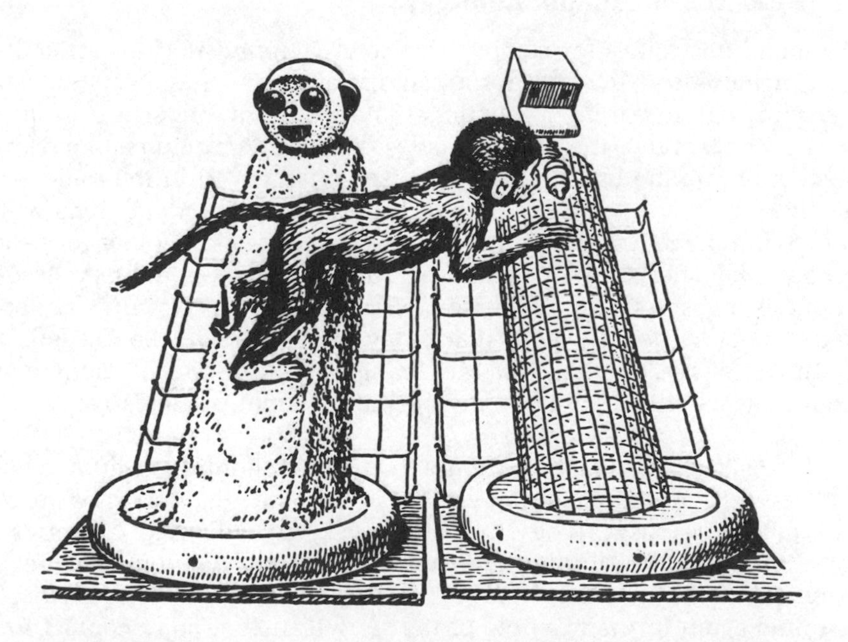

Fig. 91. A young rhesus monkey is being raised on two models: one is covered with soft terry cloth, the other is of bare wire. Below the head of the wire "mother" is a nipple dispensing milk. The young animal spends almost all of its time on the cloth "mother" and visits the wire "mother" only briefly for feeding. This is an indication that the needs for food and contact comfort are separate and can be satisfied by *different* objects. Hence, the bond does not, as was generally assumed prior to Harlow's work, depend solely on food reinforcement (after Harlow 1958).

in the comparisons between children raised in families and those raised in orphanages. This is true for specific developmental periods that, although less rigidly defined, exist also in humans. The same applies to the development of behavior aberrations as a result of a too-early separation from the mother. These effects are called the DEPRIVATION SYNDROME (lack of interest in surroundings, loss of overall activity, stereotyped movements, inability to participate in normal social behavior, etc.) (see Section 7.4.7.2).

It appears that in these behavior characteristics, especially the specificity of development during certain phases and the permanence of early experience, we have examples from animal studies that are indicative of similarities in human behavior.

More generally and with greater caution, conclusions from studies of animals not closely related to humans—such as nonprimate species—can be examined. In spite of these limitations, they have been of heuristic value in the study of similar phenomena in humans, and they have helped in the development of research strategies. Examples exist from such varied research areas as vocalizations, play behavior, social inhibitions (e.g., incest taboo), and infanticide.

12.3.2. Research in Human Ethology

A second possibility for comparative work is found in the application of research methods and by considering comparative and phylogenetic aspects in psychological research of humans. The point of departure of human ethology is the fact that the characteristics of present-day humans have come about according to the laws of evolution (see Chapter 10) in the same way as in all other organisms, with their multitude of specific adaptations to their natural environment. This invites us to search in human behavior for general, contiguous characteristics to determine their possible hereditary bases, to trace their changes in the evolution to civilized man, and to search for "earlier adaptations" in human behavior that today may no longer be needed. They may even be of disadvantage in the changed way of life of modern man, although they are still with us as phylogenetic remnants to this very day (see Section 10.5).

Such considerations can contribute to an understanding of human behavior that has hitherto been unintelligible, since the distinction is made in biology, but not in psychology, between PROXIMAL and DISTAL factors (see Sections 8.2.2.4 and 8.4.7). In other words, ethology distinguishes between phylogenetic causes and the actual regulating mechanisms of behavior. This is illustrated again in the example of the significance of body contact for the mother–child bond: young primates are not able to keep up with their troop and require their mother as a means of transport; they are in great danger when they are separated from her. It is certain that there exists a great selection pressure for the evolution of mechanisms that insure continuous physical closeness. These include behavior patterns of the young (following behavior, clinging), as well as a concomitant motivation that one may call a BONDING

DRIVE (see Section 8.3.1). This motivation has been preserved to this day, although the conditions that prevailed when this mechanism evolved in humans no longer exist in our changed way of life to the same extent as they did then. If this bonding drive finds no expression, behavior aberrations ensue that at first seem to defy our understanding. From the biological perspective, however, it is one more example for the conservative nature of preprogrammed behavior (see Section 7.1).

As a research strategy, human ethology can do comparisons between the various cultures similar to the comparative studies on animal species and populations. This strategy enables the investigator to discover behavioral characteristics that are either similar or different in the respective cultures, and that may be comparable to "species characteristics" if found in animals. This assumption seems justified, especially when the behavior occurs in comparatively "primitive" peoples and/or in cultures that are geographically widely separated and that have no communication with one another.

Another especially useful method for the investigation of the relative contribution of possibly inherited preprogrammed versus environmental influences is available in the studies of newborn infants who had little or no opportunity for learning. Further examples are people who have lost the use of individual sense modalities. This involves research with blind, deaf, and mute children. The important question here is which behavior patterns nevertheless occur normally and spontaneously, i.e., which require no specific experience, and which would be classified with innate behavior in animals (see Section 6.1.2).

12.4. Phylogenetic Preprogramming of Human Behavior

12.4.1. Examples

After the general discussion in previous chapters, we will now give some specific examples of behavior patterns with an inherited basis, which are the result of phylogenetic adaptations. These are movement coordinations, the recognition of stimuli, special learning dispositions, and motivations.

The most easily recognized examples of inherited preprogramming are movement coordinations. They are very simple movements found usually in newborn infants, e.g., the automatic head-turning movements in search of the nipple on the mother's breast; the synchronization of breathing and swallowing when nursing; the walking movements of the legs when the baby is held just above, and barely touching, the surface. Perhaps the best-known examples are the grasping reflexes, an orderly sequence of finger and toe movements that enable a newborn child to hold onto a finger or a clothesline (see Figure 92). Such behavior patterns are seen especially in prematurely born infants.

There is also a behavior pattern that does not owe its appearance to conditioning, and it appears also in deaf and mute children who had, of course, no opportunity for prior experience: smiling. It seems to be preprogrammed, and the expression of smiling is probably the best illustration of a

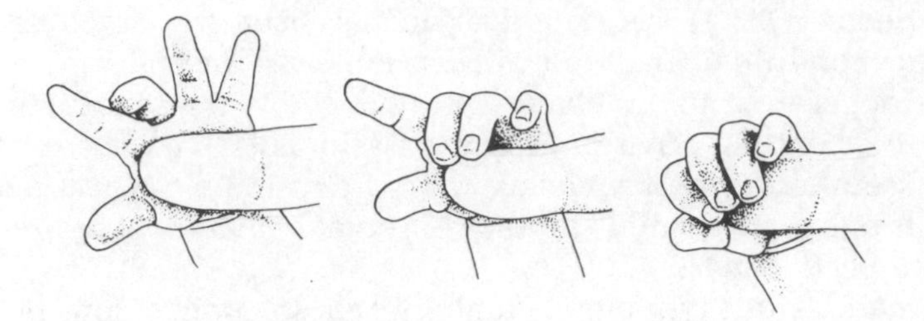

Fig. 92. The clinging, or grasping, reflex in infants (after Prechtl 1953).

modal action pattern (see Section 4.3). Additional social behavior patterns, e.g., the rapid lifting of the eyebrows, are unlikely to be explained without reference to preprogramming because of their wide distribution in human cultures. Finally, there are indications that certain behavior patterns that occur in sequences are not random but express measurable regularities that are similar to the reaction chains in animals (see Section 4.1).

Less certain is our knowledge in ethology about inherited perceptual mechanisms like the innate releasing mechanisms in animals (see Section 6.1.3). A famous example is the BABY SCHEMA, a combination of body characteristics that release a positive response in humans that can best be interpreted as a caretaking reaction with definite, appropriate emotional overtones (see Figure 93). As long as we do not have investigations that have determined the relative contribution of experience, it is impossible to conclude an inherited basis for this reaction. The same is true for the possibly preprogrammed reactions to secondary sexual characteristics of men and women.

Learning dispositions that have been discovered in various animal species (see Section 7.3) have often been suspected in man. This applies especially to the ability to learn speech, which requires much more complex muscular coordinations than many other movements, which occur much later in life than learning to speak.

There are more obvious indicators for possible inherited limits of learning abilities with respect to temporal factors: certain experiences leave more or less permanent impressions at various ages. Such a phase specificity has already been discussed in reference to socialization early in life (see Section 12.3.1).

To what degree motivations or drives, such as for aggression, may be preprogrammed in man cannot be decided on the basis of currently available data. For reasons discussed in Section 8.1.9.6 in reference to the possible biological function of aggressive motivation, such an innate basis cannot be dismissed without further study.

12.4.2. Research in Behavior Genetics in Humans

We have already pointed out the difficulties in recognizing innate behavior components (see Section 6.1.2); they are even more so with respect to human behavior. Here the inheritance of a specific behavior characteristic would be a convincing example. Hence, behavior genetic investigations have increasingly gained in popularity. Family and twin studies, complemented by investigations of adoptions, have demonstrated the existence of family-specific characteristics, especially in the cognitive area.[1] The largest number of such studies comes from psychopathological research because of the expected insights, predictability, and therapy value of such work. Relationships between chromosomal abnormalities (e.g., the doubling of the male sex Y chromosome)

[1] In psychology, COGNITIVE BEHAVIOR refers to all of those behaviors by which an individual acquires knowledge about the environment or himself. This includes perception, learning, memory, language, thinking, and making judgments. Recently, this concept also found its way into ethology, especially in primate research. In human ethology, the concept is contrasted with the more emotional aspects of behavior.

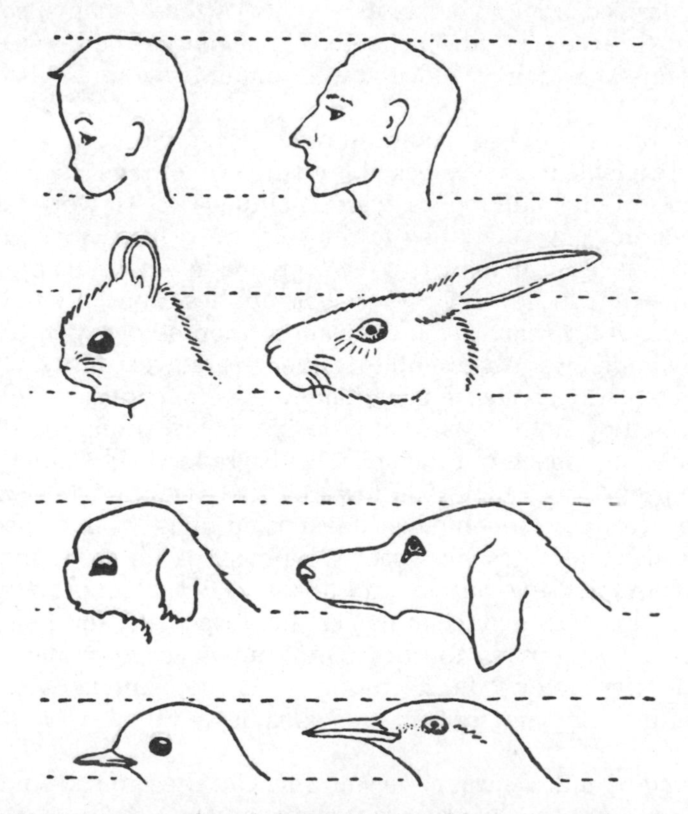

Fig. 93. Infants possess a number of characteristics and body proportions (e.g., large eyes, a high, domed forehead, a large head in proportion to the body, round cheeks) that usually elicit a positive response in adults. The same positive reaction is also transferred to young animals (second row, left), breeds of domesticated animals (third row, left), and even wild animals (European robin, fourth row, left) as long as they show the same characteristics. The reaction to this combination of stimuli, called the BABY SCHEMA, has often been cited as an example of innate recognition in humans. Whether or not this reaction is actually inborn, i.e., occurs independently of experience, can only be decided on the basis of quantitative investigations on subjects with various degrees of early experience (after Lorenz 1943 from Tinbergen 1952).

and certain deviant behavior (e.g., tendency to criminal behavior) have been discovered. There is also sex-linked inheritance in schizophrenics and alcoholics. Overall, behavior-genetic investigations have brought us insights beyond their immediate significance for therapy. They have brought preprogrammed human behavior into the realm of scientific investigation, at least as far as psychopathology is concerned. They should lay to rest the old arguments over the presence or absence of inheritance of behavior.

12.5. Prospects for the Future

This chapter was meant to show at which level conclusions based on ethological data can be made to help us understand the behavior of man. We pointed out the limits of such comparisons and suggested possible benefits that may be derived from a comparative study of the human species. It was also intended to encourage the increased collaboration between ethologists and psychologists, which is already well under way in English-speaking countries.

Two final remarks of caution are in order. First, we have only discussed the presumed benefits to some areas of human psychology if the phylogenetic considerations of the ethologists were included in its work—and if the methods of ethology were applied to the solution of particular problems by using appropriate animal models. However, it should be emphasized that positive influences can flow both ways. This applies especially to those areas in human psychology that are more advanced than ethology in their knowledge and methodology. An example is the greater variability of behavior, which is undoubtedly greater in man than in any other species. This has been largely neglected by ethologists, who were in the past primarily interested in the species-specific aspects of behavior. This is gradually changing, especially under the impact of sociobiological research (see Section 1.4). Some investigations have already documented an unexpected range of intraspecific variability in the behavior of some species. Human psychology, including its methodology in carrying out longitudinal studies, can provide a useful impetus. The same is true for cognitive studies, especially for the ontogenetic development of perceptual abilities. Great advances have been made in developmental psychology during recent years, both methodologically and with new results. In comparison, this area is relatively undeveloped in ethology.

The second remark is meant as a warning: Our present state of knowledge makes it certain that in the behavior of man there are also behaviors that originated as earlier adaptations during the course of evolution. These can be compared with similar characteristics in nonhuman organisms. Such considerations should not, however, obscure the fact that evidence available in support of this view is merely indicative and thus far does not constitute definite proof. Premature conclusions, especially those drawn on the wrong level or those which assume the false appearance of established fact, should be avoided.

References

The following list contains, first, textbooks or edited volumes covering ethology as a whole or several subareas. It is followed, in order by chapter, by references to the more specialized literature pertaining to the respective topics, which would help the reader to acquire additional and more detailed information. It must be pointed out that even in the textbooks there may be detailed treatments of specific topics that are, however, not reflected in the chapter headings because of space limitations. Edited volumes show only the name(s) of the editor(s) and not the names of individual contributors.

The references to the chapter "Ethology and Psychology" are especially numerous so that the interested reader may go to the original sources and form his own opinion, based on the diverse results, interpretations, and viewpoints that are expressed.

Many of these publications contain comprehensive references, which themselves facilitate further study.

German papers listed in the original German edition of this volume have been retained. English-language original editions have been substituted for the German translations cited in the German edition wherever possible. For German books available in English translation, only the latter has been listed. The list has been augmented by works published since the last German edition appeared.

Alcock, J., 1975: *Animal Behavior, An Evolutionary Approach.* Sinauer, Sunderland, Massachusetts.

Apfelbach, R., and Döhl, J., 1978: *Verhaltensforschung. Eine Einführung.* Fischer, Stuttgart.

Baerends, G., Beer, C., and Manning, A. (editors), 1975: *Function and Evolution in Behaviour.* Clarendon Press, Oxford.

Bateson, P. P. G., and Hinde, R. A. (editors), 1976: *Growing Points in Ethology.* Cambridge University Press, Cambridge.

Curio, E., 1976: *The Ethology of Predation.* Springer, Berlin–Heidelberg–New York.

Dawkins, R., 1976: *The Selfish Gene.* Oxford University Press, Oxford.

Dewsbury, D. A., 1978: *Comparative Animal Behavior.* McGraw-Hill, New York.

Eibl-Eibesfeldt, I., 1975: *Ethology—The Biology of Behavior,* 2nd ed. Holt, Rinehart & Winston, New York.

Ewer, R. F., 1968: *Ethology of Mammals.* Plenum Press, New York.

von Frisch, K., 1953: *The Dancing Bees.* Harcourt Brace Jovanovich, New York.

Hediger, H., 1964: *Wild Animals in Captivity.* Dover, New York.

Hediger, H., 1968: *The Psychology and Behavior of Animals in Zoos and Circuses.* Butterworths, London (Dover paperback reprint).

Hediger, H., 1968: *Man and Animal in the Zoo.* Delacorte Press, New York.

Heymer, A., 1977: *Ethological Dictionary.* Paul Parey, Berlin–Hamburg.

Hinde, R. A., 1966: *Animal Behavior: A Synthesis of Ethology and Comparative Psychology,* 2nd ed. McGraw-Hill, New York.

Holland, J. G., and Skinner, B. F., 1961: *The Analysis of Behavior*. McGraw-Hill, New York.
Immelmann, K., 1975: *Wörterbuch der Verhaltensforschung*. Kindler, Munich.
Immelmann, K. (editor), 1977: *Encyclopedia of Ethology*. Van Nostrand Reinhold, New York.
Jürgens, W., and Ploog, D., 1974: *Von der Ethologie zur Psychologie*. Kindler, Munich.
Lamprecht, J., 1974: *Verhalten*, 3rd ed. Herder, Freiburg–Basel–Vienna.
Lehner, P. N., 1979: *Handbook of Ethological Methods*, Garland STPM Press, New York–London.
Leyhausen, P., 1965: On the function of the relative hierarchy of moods (as exemplified by the phylogenetic and ontogenetic development of prey-catching in carnivores). Reprinted in: Lorenz, K., and Leyhausen, P., 1973: *Motivation of Human and Animal Behavior*. Van Nostrand Reinhold, New York.
Lorenz, K., 1970: *Studies in Animal and Human Behavior*, Vol. 1. Harvard University Press, Cambridge, Massachusetts.
Lorenz, K., 1971: *Studies in Animal and Human Behavior*, Vol. 2. Harvard University Press, Cambridge, Massachusetts.
Lorenz, K., 1978: *Vergleichende Verhaltensforschung. Grundlagen der Ethologie*. Springer, Vienna–New York.
Manning, A., 1979: *An Introduction to Animal Behavior*, 3rd ed. Addison-Wesley, Reading, Massachusetts.
Marler, P. R., and Hamilton, W. J., III, 1966: *Mechanisms of Animal Behavior*. Wiley, New York.
McGill, T. E. (editor), 1977: *Readings in Animal Behavior*, 3rd ed. Holt, Rinehart & Winston, New York.
Ploog, D., and Gottwald, P., 1974: *Verhaltensforschung. Instinkt–Lernen–Hirnfunktion*. Urban und Schwarzenberg, Munich.
Schiller, C. H., 1957: *Instinctive Behavior*. International Universities Press, New York.
Skinner, B. F., 1966: *The Behavior of Organisms. An Experimental Analysis*. Appleton-Century-Crofts, New York.
Stokes, A. W., 1968: *Animal Behavior in Laboratory and Field*. W. H. Freeman, San Francisco–London.
Tembrock, G., 1978: Verhaltensbiologie. In: *Wörterbücher der Biologie*. Fischer, Jena.
Thorpe, W. M., 1963: *Learning and Instinct in Animals*, 2nd ed. Methuen, London.
Tinbergen, N., 1948: Social releasers and the experimental method required for their study. *Wilson Bulletin* 60, 6–52.
Tinbergen, N., 1951 (paperback 1969): *The Study of Instinct*. Oxford University Press (paperback), Oxford.
Wickler, W., and Seibt, U. (editors), 1973: *Vergleichende Verhaltensforschung*. Hoffman und Campe, Hamburg.
Wickler, W., and Seibt, U., 1977: *Das Prinzip Eigennutz*. Hoffman und Campe, Hamburg.
Wilson, E. O., 1975: *Sociobiology: The New Synthesis*. Belknap Press, Cambridge, Massachusetts.

Basic Ethological Concepts

Becker-Carus, C., *et al.*, 1972: Motivation, Handlungsbereitschaft, Trieb. *Z. Tierpsychol.* **30,** 321–326.
Heiligenberg, W., 1964: Ein Versuch zur ganzheitsbezogenen Analyse des Instinktverhaltens eines Fisches (*Pelmatochromis subocellatus kribensis* Boul., Cichlidae). *Z. Tierpsychol.* **21,** 1–52.
Lorenz, K., 1937: Über den Begriff der Instinkthandlung. *Folia Biotheoretica* **II,** Series B, 17–50.
Pawlow, J. P., 1941: *Lectures on Conditioned Reflexes*, 2 vols. International Publishers, New York.

External Stimuli

Curio, E., 1969: Der Funktionszusammenhang zwischen einer Handlung und der ihr zugrundeliegenden Erregung als Grundlage der Ethometrie von Schlüsselreizen. *Z. Vgl. Physiol.* **62,** 301–317.

Schleidt, W. M., 1962: Die historische Entwicklung der Begriffe "Angeborenes auslösendes Schema" und "Angeborener Auslösemechanismus" in der Ethologie. *Z. Tierpsychol.* **19,** 697–722.

Schleidt, W. M., 1964: Wirkungen äusserer Faktoren auf das Verhalten. *Fortschr. Zool.* **16,** 469–499.

Tinbergen, N., and Perdeck, A. C., 1951: On the stimulus situation releasing the begging response in the newly-hatched herring gull chick *(Larus argentatus)*. *Behaviour* **3,** 1–38.

Temporal and Hierarchical Organization of Behavior

Andrew, R. J., 1957: The aggressive and courtship behaviour of certain Emberizinae. *Behaviour* **10,** 255–308.

Baerends, G. P., 1956: Aufbau tierischen Verhaltens. In: Kükenthal, *Handbuch der Zoologie* 8, **10**(3), 1–32.

Baerends, G. P., 1976: The functional organization of behaviour. *Anim. Behav.* **24,** 726–738.

von Holst, E., and von St. Paul, V., 1963: On the functional organization of drives. *Anim. Behav.* **XI,** 1.

Pattee, H. H., 1973: *Hierarchy Theory: The Challenge of Complex Systems.* Braziller, New York.

Sevenster, P., 1961: A causal analysis of a displacement activity: Fanning in *Gasterosteus aculeatus. Behaviour*, Suppl. No. **9**.

Tinbergen, N., 1950: The hierarchical organization of nervous mechanisms underlying instinctive behaviour. *Symp. Soc. Exp. Biol.* **4,** 305–312.

Behavioral Physiology

Beach, F. A., 1961: *Hormones and Behavior.* Hoeber, New York.

Dörner, G., 1972: *Sexualhormonabhängige Gehirndifferenzierung und Sexualität.* Springer, Vienna and New York.

Ewert, J.-P., 1980: *Neuro-Ethology.* Springer, Berlin–Heidelberg–New York.

Fentress, J. C., 1976: *Simpler Networks and Behavior.* Sinauer, Sunderland, Massachusetts.

Hess, W. R., 1949: *Das Zwischenhirn.* Schwabe, Basel.

von Holst, E., 1970: *Zur Verhaltensphysiologie bei Tieren und Menschen*, Vol. II. Piper, Munich.

von Holst, E., 1973: *The Behavioural Physiology of Animals and Man*, Vol. I. University of Miami Press, Coral Gables, Florida.

Milner, P. M., 1971: *Physiological Psychology.* Holt, Rinehart & Winston, London.

Neumann, F., and Steinbeck, H., 1971: Hormonelle Beeinflussung des Verhaltens. *Klin. Wochenschr.* **49,** 790–806.

Roeder, K. W., 1963: *Nerve Cells and Insect Behavior.* Harvard University Press, Cambridge, Massachusetts.

Wright, P., Caryl, P. G., and Vowles, D. M. (editors), 1975: *Neural and Endocrine Aspects of Behaviour in Birds.* Elsevier, Amsterdam–Oxford–New York.

Ontogeny of Behavior

Burghardt, G., and Bekoff, M., 1978: *Comparative and Evolutionary Aspects of Behavioral Development.* Garland, New York.

Gottlieb, G., 1971: *Development of Species Identification in Birds.* University of Chicago Press, Chicago–London.

Kuo, Z. Y., 1967: *The Dynamics of Behavior Development.* Random House, New York.

Lorenz, K., 1965: *Evolution and Modification of Behavior.* University of Chicago Press, Chicago.

Moltz, H., 1971: *The Ontogeny of Vertebrate Behaviour.* Academic Press, New York–London.

Learning

Angermeier, W. F., 1976: *Kontrolle des Verhaltens. Das Lernen als Erfolg.* Springer, Berlin–Heidelberg–New York.

Buchholtz, C., 1973: *Das Lernen bei Tieren.* Fischer, Stuttgart.

Foppa, K., 1965: *Lernen, Gedächtnis, Verhalten.* Kiepenheuer und Witsch, Cologne–Berlin.

Hess, H. H., 1973: *Imprinting.* Van Nostrand Reinhold, New York.

Hinde, R. A., and Stevenson-Hinde, J., 1973: *Constraints on Learning.* Academic Press, London–New York.

Köhler, W., 1925: *The Mentality of Apes.* Harcourt Brace, New York.

Lauer, J., and Lindauer, M., 1971: *Genetisch fixierte Lerndispositionen bei der Honigbiene.* Steiner, Wiesbaden.

Rensch, B., 1973: *Gedächtnis, Begriffsbildung und Planhandlungen bei Tieren.* Parey, Berlin–Hamburg.

Sluckin, W., 1973: *Imprinting and Early Learning.* Aldine, Chicago.

Tschanz, B., 1968: *Trottellummen.* Suppl. **4** to *Z. Tierpsychol.*

Social Behavior

Bastock, M., 1967: *Courtship—An Ethological Study.* Aldine, Chicago.

Crook, J. H. (editor), 1970: *Social Behaviour in Birds and Mammals.* Academic Press, London–New York.

Hendrichs, H. H., 1978: Die soziale Organisation von Säugetierpopulationen. *Säugetierkundl. Mitt.* **26,** 81–116.

Hinde, R. A. (editor), 1972: *Non-verbal Communication.* Cambridge University Press, Cambridge, Massachusetts.

Hutchinson, J. B. (editor), 1978: *Biological Determinants of Sexual Behaviour,* Wiley, London.

Jenni, D. A., 1974: Evolution of polyandry in birds. *Am. Zool.* **14,** 129–144.

Koehler, O., 1969: Tier- und Menschensprachen. In: Altner, G. (editor): *Kreatur Mensch.* Moos, Munich.

Kummer, H., 1971: *Primate Societies.* Aldine, Chicago.

Lorenz, K., 1966: *On Aggression.* Harcourt, Brace and World, New York.

Nice, M. M., 1962: Development of behavior in precocial birds. *Trans. Linn. Soc. N.Y.* **VIII.**

Nicolai, J., 1970: *Elternbeziehung und Partnerwahl im Leben der Vögel.* Piper, Munich.

Remane, A., 1971: *Sozialleben der Tiere.* Fischer, Stuttgart.

Schenkel, R., 1956 and 1958: Zur Deutung der Balzleistungen einiger Phasianiden und Tetraoniden. *Ornithol. Beob.* **53,** 182–201, and **55,** 65–95.

Sebeok, T. A. (editor), 1977: *How Animals Communicate.* Indiana University Press, Bloomington–London.

Skutch, A. F., 1976: *Parent Birds and their Young.* University of Texas Press, Austin–London.

Smith, W. J., 1977: *The Behavior of Communicating: An Ethological Approach.* Harvard University Press, Cambridge, Massachusetts.

Trivers, R. L., 1974: Parent–offspring conflict. *Am. Zool.* **14,** 249–264.

Behavior Genetics

Abeelen, J. H. F. (editor), 1974: *The Genetics of Behavior.* North-Holland, Amsterdam–Oxford.

De Fries, J. C., and Plomin, R., 1978: Behavioral genetics. *Ann. Rev. Psychol.* **29,** 473–515.

Ehrman, L., Omen, G. S., and Caspari, E. (editors), 1972: *Genetics, Environment, and Behavior.* Academic Press, London–New York.

Fuller, J. F., and Thompson, W. R., 1978: *Foundations of Behavior Genetics.* C. V. Mosby, St. Louis.

Halsey, A. H. (editor), 1977: *Heredity and Environment.* Methuen, London.

Phylogenetic Development of Behavior

Brown, J. L., 1975: *The Evolution of Behavior.* W. W. Norton, New York.
Gould, S. J., 1977: *Ontogeny and Phylogeny.* Belknap Press, Cambridge, Massachusetts.
Jolly, A., 1972: *The Evolution of Primate Behavior.* Macmillan, New York.
Kurth, G., and Eibl-Eibesfeldt, I. (editors), 1975: *Hominisation und Verhalten.* Fischer, Stuttgart.
Mayr, E., 1963: *Animal Species and Evolution.* Belknap Press, Cambridge, Massachusetts.
Mayr, E. 1974: Behavior programs and evolutionary strategies, *Am. Sci.* **62,** 650–659.
Roe, A., and Simpson, G. G., 1958: *Behavior and Evolution.* Yale University Press, New Haven, Connecticut.
Siewing, R. (editor), 1978: *Evolution.* Fischer, Stuttgart–New York.
Wickler, W., 1967: Vergleichende Verhaltensforschung und Phylogenetik. In: G. Heberer (editor): *Die Evolution der Organismen,* 3rd ed., Vol. 1. Fischer, Stuttgart.
Wickler, W., 1970: *Stammesgeschichte und Ritualisierung.* Piper, Munich.

Influence of Domestication on Behavior

Fraser, A. F., 1974: *Farm Animal Behavior.* Bailliere Tindall, London.
Hafer, E. S. E. (editor), 1975: *The Behaviour of Domestic Animals,* 3rd ed. Bailliere Tindall, London.
Herre, W., and Röhrs, M., 1973: *Haustiere—Zoologisch gesehen.* Fischer, Stuttgart.
Lorenz, K., 1940: Die angeborenen Formen möglicher Erfahrung. *Z. Tierpsychol.* **5,** 235–409.
Sambraus, H. H., 1978: *Nutztier-Ethologie.* Parey, Berlin–Hamburg.
Sossinka, R., 1970: Domestikationserscheinungen beim Zebrafinken *Taeniopygia guttata castanotis* (Gould), *Zool. Jahrb. Abt. Syst. Oekol. Geogr. Tier.* **97,** 455–521.

Ethology and Psychology

Bischof, N., 1972: Inzuchtbarrieren in Säugetiersozietäten. *Homo* **23,** 330–351.
Blurton-Jones, N. (editor), 1972: *Ethological Studies in Child Behaviour.* Cambridge University Press, Cambridge.
Bowlby, J., 1969: *Attachment.* Hogarth, London.
Bowlby, J., 1973: *Separation: Anxiety and Anger.* Hogarth, London.
Charlesworth, W. R., 1973: Ethology's contribution to a framework for relevant research. Occasional Papers No. 24, American Psychological Association, Montreal.
von Cranach, M. (editor), 1976: *Methods of Inference from Animal to Human Behaviour.* Aldine, Chicago–The Hague–Paris.
Eibl-Eibesfeldt, I., 1976: *Der vorprogrammierte Mensch.* Deutscher Taschenbuch-Verlag, Munich.
Eysenck, H. J., 1978: *The Nature and Measurement of Intelligence.* Springer, London.
Goodson, F. E., 1973: *The Evolutionary Foundations of Psychology.* Holt, Rinehart & Winston, New York.
Grossman, K. (editor), 1977: *Entwicklung der Lernfähigkeit.* Kindler, Munich.
Grossmann, K., and Immelmann, K., 1978: Phasen der frühkindlichen Entwicklung. In: Wendt, H. (editor): *Der Mensch,* Vol. 3. Kindler, Munich.
Harlow, H. F., Gluck, J. P., and Suomi, S. J., 1972: Generalization of behavioral data between nonhuman and human animals. *Am. Psychol.* **27,** 709–716.
Hassenstein, B., 1973: *Verhaltensbiologie des Kindes.* Piper, Munich.
Immelmann, K., 1971: Ontogenetische Entwicklung sozialer Beziehungen bei Mensch und Tier. *Naturwiss. Rundsch.* **24,** 325–334.
Kaplan, A. R. (editor), 1976: *Human Behavior Genetics,* Charles C. Thomas, Springfield, Illinois.
Klaus, M. H., and Kennell, J. H. (editors), 1976: *Maternal Infant Bonding.* C. V. Mosby, St. Louis.
Koehler, O., 1969: Tier- und Menschensprachen. In: G. Altner (editor): *Kreatur Mensch.* Moos, Munich.

Lenneberg, E. H., 1967: *Biological Foundations of Language*. Wiley & Sons, New York.
Loehlin, J. C., and Nichols, R. C., 1976: *Heredity, Environment and Personality*. Texas University Press, Austin.
McGaugh, J. (editor), 1971: *Psychology: Behavior from a Biological Perspective*. Academic Press, New York.
McGrew, W. C., 1972: *An Ethological Study of Children's Behavior*. Academic Press, New York.
Papoušek, H., and Papoušek, M., 1978: Interdisciplinary parallels in studies of early human behavior: From physical to cognitive needs, from attachment to dyadic education. *Int. J. Behav. Dev.* **1,** 37–49.
Piaget, J., and Inhelder, B., 1977: *Die Psychologie des Kindes*. Fischer, Frankfurt/Main.
Sackett, G. P. (editor), 1978: *Observing Behavior: Theory and Applications in Mental Retardation*. University Park Press, Baltimore.
Schaffer, H. R. (editor), 1977: *Studies in Mother–Infant Interaction*. Academic Press, London.
Vandenberg, S. G. (editor), 1965: *Methods and Goals in Human Behavior Genetics*, Academic Press, New York.
Washburn, S. L., 1978: Human behavior and the behavior of other animals. *Am. Psychol.* **33,** 405–418.
Wilson, E. O., 1978: *Human Nature*. Harvard University Press, Cambridge, Massachusetts.

Figure Credits

Figures 1, 2, 3, 4, 10, 12, 13, 15, 16, 17, 18, 21, 23, 24, 25, 26, 87, 93: Tinbergen, N., 1952: *Instinktlehre.* Parey, Berlin–Hamburg.

Figure 6: Romer, A. S., 1959: *Vergleichende Anatomie der Wirbeltiere.* Parey, Hamburg–Berlin.

Figure 7: Rensing, L., Hardeland, R., Runge, M., and Galling, G., 1975: *Allgemeine Biologie.* Ulmer, Stuttgart.

Figure 8: Buchholtz, C., 1971: Zur Formkonstanz des Labiumschlages der Larve von *Aeschna cyanea.* In: A. Stokes: *Praktikum der Verhaltensforschung.* Fischer, Stuttgart.

Figure 9: Curio, E., 1968: Die Adaption einer Handlung ohne den zugehörigen Bewegungsablauf. *Verh. Dtsch. Zool. Ges.* **1967,** 153–163.

Figure 11: Becker-Carus, C. *et al.*, 1972: Motivation, Handlungsbereitschaft, Trieb. *Z. Tierpsychol.* **30,** 321–326.

Figure 14: Tinbergen, N., and Perdeck, A. C., 1951: On the stimulus situation releasing the begging response in the newly-hatched herring gull chick *(Larus argentatus)*, *Behaviour* **3,** 1–38.

Figure 20: Marler, P., 1956: Über die Eigenschaften einiger tierlicher Rufe. *J. Ornithol.* **97,** 220–227.

Figure 22: Lorenz, K., and Tinbergen, N., 1938: Taxis und Instinkthandlung in der Eirollbewegung der Graugans. *Z. Tierpsychol.* **2,** 1–29.

Figure 28: Hess, W. R., 1957: Die Formatio reticularis des Hirnstammes im verhaltens-physiologischen Aspekt. *Arch. Psychiatr. Neurol.* **196,** 329–336.

Figure 29: Huber, F., 1965: Aktuelle Probleme in der Physiologie des Nervensystems der Insekten. *Naturwiss. Rundsch.* **18,** 143–156.

Figures 30, 31, 32, 33, 35: von Holst, E., and von St. Paul, U., 1960: Vom Wirkungsgefüge der Triebe. *Naturwissenschaften* **47,** 409–422.

Figure 34: von St. Paul, U., 1964: Zur Frage der hierarchischen Ordnung instinktiven Verhaltens. *Biol. Jahresh.* **4,** 105–116.

Figures 36, 69: Lehrman, D. S., 1964: The reproductive behavior of ring doves. *Sci. Am.* **211**(5), 48–54.

Figure 37: Eibl-Eibesfeldt, J., 1963: Angeborenes und im Verhalten einiger Säuger. *Z. Tierpsychol.* **20,** 705–754.

Figure 38: Hess, E., 1956: Space perception in chicks. *Sci. Am.* **195,** 71–80.

Figure 39: Hess, E., 1975: *Prägung.* Kindler, Munich.

Figure 41: Lauer, J., and Lindauer, M., 1971: *Genetisch Fixierte Lerndispositionen bei der Honigbiene.* Steiner, Wiesbaden.

Figure 43: Köhler, W., 1921: *Intelligenzprüfungen an Menschenaffen.* Springer, Berlin.

Figure 44: von Frisch, O., 1962: Zur Biologie des Zwergchamäleons. *Z. Tierpsychol.* **19,** 276–289.

Figure 48: Hess, E., 1959: Imprinting. *Science* **130,** 133–141.

Figures 49, 50: Nicolai, J., 1964: Der Brutparasitismus der Viduinae als ethologisches Problem. *Z. Tierpsychol.* **21,** 129–204.

Figures 51, 52: von Frisch, K., 1965: *Aus dem Leben der Bienen,* 8th ed. Springer, Berlin–Heidelberg–New York.

Figure 53: Eibl-Eibesfeldt, I., and Sielmann, H., 1962: Beobachtungen am Spechtfinken *Cactospiza pallida* (Sclater und Salvin). *J. Ornithol.* **103,** 92–101.

Figure 54: Kästle, W., 1963: Zur Ethologie der Grasanolis *(Norops auratus)*. *Z. Tierpsychol.* **20,** 12–33.

Figure 55: Tinbergen, N., 1959: Comparative studies of the behaviour of gulls (Laridae): A progress report. *Behaviour* **15,** 1–70.

Figure 56: Walther, F., 1965: Verhaltensstudien an der Grantgazelle (*Gazella granti* Brooke, 1872) im Ngorongoro-Krater. *Z. Tierpsychol.* **22,** 167–208.

Figure 57: Klingel, H., 1975: Die soziale Organisation der Equiden. *Verh. Dtsch. Zool. Ges.* **1975,** 71–80.

Figures 58, 60: Hediger, H., 1954: *Skizzen zu einer Tierpsychologie im Zoo und im Zirkus.* Europa, Stuttgart.

Figure 59: Heymer, A., 1973: Verhaltensstudien an Prachtlibellen. *Fortsch. Verhaltensforsch,* No. 11. Parey, Berlin.

Figure 61: Linsenmair, K. E., 1967: Konstruktion und Signalfunktion der Sandpyramide der Reiterkrabbe *Ocypode saratan* Forsk. *Z. Tierpsychol.* **24,** 403–456.

Figure 63: Walther, F., 1958: Zum Kampf- und Paarungsverhalten einiger Antilopen. *Z. Tierpsychol.* **15,** 340–380.

Figure 64: Thomas, E., 1955: Der Kommentkampf der Kreuzotter (*Vipera berus* L.). *Naturwissenschaft* **42,** 539.

Figure 65: Schöne, H., and Schöne, H., 1963: Balz und andere Verhaltensweisen der Mangrovekrabbe *Goniopsis cruentata* Latr. und das Winkverhalten der eulitoralen Brachyuren. *Z. Tierpsychol.* **20,** 641–656.

Figure 66: Sick, H., 1959: Die Balz der Schmuckvögel. *J. Ornithol.* **100,** 269–302.

Figure 67: Gwinner, E., 1964: Untersuchungen über das Ausdrucks- und Sozialverhalten des Kolkraben (*Corvus corax corax* L.). *Z. Tierpsychol.* **21,** 657–748.

Figure 68: Nicolai, J., 1956: Zur Biologie und Ethologie des Gimpels. *Z. Tierpsychol.* **13,** 93–132.

Figure 70: Wagner, H. O., 1954: Versuch einer Analyse der Kolibribalz. *Z. Tierpsychol.* **11,** 182–212.

Figure 71: Jacobs, W., 1966: Die Gesänge der Heuschrecken. In: Burkhardt, D., Schleidt, W., and Altner, H.: *Signale in der Tierwelt.* Moos, Munich.

Figures 76, 88: Tinbergen, N., 1955: *Tiere untereinander.* Parey, Berlin–Hamburg.

Figure 77: McBride, G., 1971: *Animal Families.* Reader's Digest Association, Pleasantville, New York.

Figure 78: Din, N. A., and Eltringham, K., 1974: Ecological separation between white and pinkbacked pelicans in the Ruwenzori National Park, Uganda. *Ibis, J. Br. Ornithol. Union* **116,** 28–43.

Figure 79: Wickler, W., 1967: Vergleichende Verhaltensforschung und Phylogenetik. In: Heberer, G.: *Die Evolution der Organismen,* 3rd ed. Fischer, Stuttgart.

Figure 80: Pilleri, G., and Knuckey, J., 1969: Behaviour patterns of some Delphinidae observed in the western Mediterranean. *Z. Tierpsychol.* **26,** 48–72.

Figure 81: Wickler, W., 1969: Sind wir Sünder—Naturgesetze der Ehe. Droemer, Munich.

Figure 82: Stadie, C., 1968: Verhaltensweisen von Gattungsbastarden *Phasianus colchicus* × *Gallus gallus* f. domestica in Vergleich mit denen der Ausgangsarten. *Verh. Dtsch. Zool. Ges.* **1967,** 493–510.

Figure 83: Dilger, W., 1962: Methods and objectives of ethology. *Living Bird* **1,** 83–92.

Figure 84: Osche, G., 1966: *Die Welt der Parasiten.* Springer, Berlin–Heidelberg–New York.

Figure 85: Right: Steiner, H., 1955: Das Brutverhalten der Prachtfinken, Spermestidae, als Ausdruck ihres selbständigen Familiencharakters. *Acta XI Congr. Int. Ornithol.* Basel 1954, 350–355.

Figure 86: Thorpe, W. H., and P. M. Pilcher, 1958: The nature and characteristics of subsong. *Br. Birds* **51,** 509–514.

Figure 90: Sossinka, R. 1972: Domestikationserscheinungen beim Zebrafinken *Taeniopygia guttata castanotis* (Gould). *Zool. Jahrb. Abt. Syst. Oekol. Geogr. Tier.* **97,** 455–521.

Figure 91: Harlow, H. F., 1958: The nature of love. *Am. Psychol.* **13,** 673–685.

Figure 92: Prechtl, H. F. R., 1953: Stammesgeschichtliche Reste im Verhalten des Säuglings. *Umschau* **53,** 656–658.

Author Index

Animal Index

Subject Index